What Information Do I Need From My Provider?

Local Exchange Carrier

☐ Contact information
Phone company rep: _____
Company name: _____
Phone: _____

☐
☐ Installation date: _____
Service order number: _____

ISDN service charges
☐
☐ Monthly charge $_____
Per minute usage
8 a.m. to 5 p.m. $_____
from _____ to _____ $_____
☐ Additional charges $_____

☐ Installation information
Phone company rep: _____
Company name: _____
Phone: _____

Provisioning information
☐ Type of ISDN
☐ National
☐ Custom

☐ Digital switch type
☐ 5ESS
☐ DMS-100
☐ Siemens EWSD

☐ Local directory numbers
☐ B channel one   _____
☐ B channel two   _____

☐ Service profile identifiers (SPIDs)
☐ B channel one   _____
☐ B channel two   _____

☐ B channels
☐ Voice
☐ Data
☐ Both

ISDN Equipment Vendor Information

☐ ISDN Ordering Code (IOC) _____

Mastering™ ISDN

Mike Sapien
Greg Piedmo

SYBEX®

San Francisco • Paris • Düsseldorf • Soest

Associate Publisher: Gary Masters
Acquisitions Manager: Kristine Plachy
Acquiring Editor: John Read
Editor: Marilyn Smith
Project Editor: Ben Miller
Technical Editor: Robert Blackshaw
Book Designer: Catalin Dulfu
Graphic Illustrator: Patrick Dintino
Electronic Publishing Specialist: Dina Quan
Production Coordinator: Robin Kibby
Indexer: Nancy Guenther
Cover Designer: Design Site

SYBEX is a registered trademark of SYBEX Inc.
Mastering is a trademark of SYBEX Inc.

TRADEMARKS: SYBEX has attempted throughout this book to distinguish proprietary trademarks from descriptive terms by following the capitalization style used by the manufacturer.

Netscape Communications, the Netscape Communications logo, Netscape, and Netscape Navigator are trademarks of Netscape Communications Corporation.

The author and publisher have made their best efforts to prepare this book, and the content is based upon final release software whenever possible. Portions of the manuscript may be based upon pre-release versions supplied by software manufacturer(s). The author and the publisher make no representation or warranties of any kind with regard to the completeness or accuracy of the contents herein and accept no liability of any kind including but not limited to performance, merchantability, fitness for any particular purpose, or any losses or damages of any kind caused or alleged to be caused directly or indirectly from this book.

Library of Congress Card Number: 96-69666
ISBN: 0-7821-1845-3

Manufactured in the United States of America

10 9 8 7 6 5 4 3 2 1

To Debbie, Nicole, and Brendan,
my wife, daughter, and son
Mike Sapien

To the loving memory of my brother Gary
Greg Piedmo

ACKNOWLEDGMENTS

I would like to thank Marilyn Smith as the great editor, author, motivator, and organizer who helped make this book come together and keep moving. At Sybex, I would like to thank Kristine Plachy for helping pull the contract issues through, John Read for insisting that I needed to do this book, and Ben Miller for his production support. Also, thanks to Victor Romero for referring Sybex to me for this book.

Greg Piedmo, my co-author, quickly jumped in and added greatly to the energy and authorship of this book. Greg was great to work with and share the stress with during this writing.

At Bellcore, Karen FitzGerald helped kick off this book and helped tremendously on the international content. I would also like to thank Royce Hazard-Leonards for all the research and background information, including her personal experience and involvement with ISDN.

My appreciation also goes to all the personnel at Ascend Communications, Shiva Corporation, Motorola, Adtran, and Cisco for all their support and the product and industry information they supplied.

I would like to thank the *ISDN News* staff, particularly Lori Marie Sylvia, for their responsiveness and the information they shared about the industry.

At IGICOM, Molly Schaeffer helped by supplying information about Global 96 events. Thanks also go to Kathie Blankenship from Smart Valley Group for her help in supplying the telecommuting application and survey information.

Mike Sapien, September 1996

My thanks to Dr. Fred Saba for his mentorship and his encouragement, guidance, and support. He makes the impossible sound possible.

Special thanks to the folks at the San Diego State University's Educational Technology Department, including Pam Monroe for her support and Bob Hoffman, Ph.D, Donn Ritchie, Ph.D, Brockenbrough S. Allen, Ph.D, Bernie Dodge, Ph.D, and Alison Rossett, Ph.D, for their inspiration and pushing me to be my best.

Thanks also go to the research fellows working for Pacific Bell's Education First initiative. A special thanks to Linda Hyman.

I would also like to thank Mike Sapien, my co-author, who took a chance on a relative unknown and opened a door of opportunity for me.

Thanks go to the folks at Sybex: Kristine Plachy and John Read for providing the opportunity to do this book, and Ben Miller for his patience and perseverance. Thanks to Marilyn Smith, our editor, counselor, coach, and cheerleader. Thanks also go to the illustrator, Pat Dintino, the electronic publishing specialist, Dina Quan, and the production coordinator, Robin Kibby, for all their hard work in producing this book.

Thanks to my Mom and Dad for years of support and love that provided the sustenance for a journey that has brought me to this point.

And thanks to my friends for encouragement and understanding: Gail Lucas, Dave and Jennifer Lewis, Diane and Dallas Jones, Roger McElmell, Rusty Meike, and especially Lori McQuillen, who at least tried to get me organized.

I also wish to express my appreciation for the help I got from the equipment vendor personnel. Thanks to 3Com's Jennifer Wade and to Elizabeth Cardinali, who always followed up on requests and were fast and friendly. Special thanks to Brian Patnoe at Motorola for his service, support, and bending over backward to help. Thanks to

Kevin Flanagan and Patricia Chaplin at PictureTel; Vickie Pierce, her secretary Katy, and Kathy Melcher at Intel; Mia Bradway at Compression Labs Inc.; and Mike Beltrano at US Robotics. Special thanks to the guys at American Digital Network for their service and support.

Greg Piedmo, September 1996

CONTENTS AT A GLANCE

TABLE OF CONTENTS

6   About Telecommuting Applications   201

PART III   How Do I Get Connected?   289

9   How Do I Get Started?   291

10 A Guide to ISDN Setup for Online Information Access

11 A Guide to ISDN Setup for Telecommuting and RLA 363

INTRODUCTION

If you're reading this, we can assume that you've heard something about ISDN. *Mastering ISDN* provides the background and information that you need to move from an initial awareness of some new service called ISDN and into the knowledgeable group of people who know how to put the service to practical use and why.

ISDN (Integrated Services Digital Network) is a high-speed, digital communication service that is an alternative to standard analog telephone service. ISDN has been widely adopted in Europe, but is just now gaining acceptance and becoming more available in the United States. Much has been written about the technical details of ISDN, but there is little published to help people actually use the service. This book is a "nuts and bolts" guide for those who want to get connected and use ISDN. You'll find all the information you need for getting your ISDN equipment, setting up, and ordering ISDN service, with specific instructions for your particular application.

Who Should Read This Book

You may want to read this book because you are interested in learning about any new technology that comes along. But more likely, you are one of the many people who do not have an advanced technical degree or particular interest in telecommunications, but are being pushed into learning more because of something you (or your employer) bought or are thinking of buying. The new online service you subscribed to, the video calls you want to make to your relatives using your new PC, or that LAN remote access package you need for work may be the reason that you now have an interest in ISDN. You may be an MIS manager who is considering using ISDN service for your LAN-to-LAN and remote-access connections. Or maybe you've noticed that ISDN is starting to be packaged in services and in PC hardware and software, and that many Internet service providers are suggesting ISDN for better graphics capabilities and faster access to the World Wide Web. Now you're beginning to think that there is something better than the fast modem technology that you just mastered.

If you are not one of those lucky people who have a friend who is an expert on computer networks, or a family member who is a telecommunications technician for the phone company, this book will become that friend. It puts the inside knowledge of this new digital telecommunications service and how to use it at your fingertips.

In this book, you'll learn how ISDN service can make your time on the PC more productive and greatly improve the speed with which you can connect and access information from your PC at work, at home, or on the road.

How This Book Is Organized

This book is organized in four parts:

- *Part I: What Is ISDN?* describes this digital telecommunications service. The first chapter provides some background on where ISDN came from and where it is now. Chapter 2 explains why ISDN may be your best choice today. Chapter 3 gives all the details about the makeup of ISDN. The final chapter in this part explains how ISDN has developed in countries other than the United States and about your international ISDN connections.

- *Part II: What Can I Do with ISDN?* provides information about the four main applications of ISDN. In Chapters 5 through 8, you will learn about ISDN and online information access, telecommuting, desktop videoconferencing, and remote LAN access. These chapters include real-life case studies of the applications in action.

- *Part III: How Do I Get Connected?* is your guide to setting up and ordering ISDN service for your particular application. Each chapter in this part, Chapters 9 through 12, includes an ISDN worksheet, equipment recommendations, and detailed instructions for the applications covered in Part 2.

- *Part IV: Issues and Trends* takes you to the next stage with your ISDN service. Chapter 13 covers management issues that may be of concern to those in charge of their company's ISDN service users. Chapter 14 describes the most common problems that may come up with setting up ISDN service and offers some suggestions on how to avoid them. The subject of the final chapter in the book is the future of ISDN, including trends and new applications.

The book also includes three appendices that point to a variety of resources for learning more about ISDN. Appendix A lists contact information for ISDN carriers. In Appendix B, you'll find a list of ISDN equipment vendors and their contact information. Appendix C includes all the other types of ISDN resources available, ranging from ISDN user groups to Web pages that address specific applications for ISDN. At the back of the book you'll find a comprehensive glossary. Turn there whenever you come across an ISDN or another telecommunications term that isn't familiar to you.

Throughout the book you will find notes, tips, and warnings, which serve as helpful adjuncts to the main text.

If you're beginning to use ISDN technology, you're stepping into the digital world of high-speed telecommunications, with all its advantages over the old analog technology. We hope this book helps you along your way. If you would like to contact the authors, you can send e-mail to Mike Sapien at sapien3@aol.com and to Greg Piedmo at piedmo@rmi.net.

PART I

What Is ISDN

CHAPTER
ONE

1

ISDN Defined

- ■ Features of the ISDN communications service

- ■ How ISDN evolved

- ■ Advances in ISDN hardware and service

- ■ Where ISDN is today

Unfortunately, ISDN's name does not clearly identify the type of service or communication that it is nor the functions that it provides to the user. ISDN stands for Integrated Services Digital Network. It is a relatively new type of communications method. In this chapter, we'll give you a clearer idea of just what ISDN is, how it has evolved, and where the technology is today.

ISDN's Fast, Reliable Service and Multiple Connections

ISDN is a *digital* communications option that is different than your regular phone service today. Your regular phone service uses older *analog* technology.

The analog signals used for the regular phone service are transmitted as sound waves traveling on the copper wire. In the analog world, signals amplified over long distances carry all the noise and distortion they accumulate along the way. The digital signals of ISDN are in the form of pulses of ones and zeros, with error correction and recovery methods that provide more reliable and accurate transmissions. With digital transmission, the ones and zeros are detected and regenerated as new, clean signals, without any noise or distortion passed on. Chapter 2 provides more details about the differences between digital and analog communications service.

Two main types of ISDN are available:

- BRI, or Basic Rate ISDN, is the most appropriate type of ISDN service for personal computer connections and most business and individual uses.

- PRI, or Primary Rate ISDN, is designed for applications that need to transmit larger amounts of data, such as in LAN interconnections, PBX (Private Branch Exchange) connections, and videoconferencing applications.

Basic Rate ISDN provides three communications channels on the same line for its service: two B (or bearer) channels and one D (or signal) channel. The two B channels can be used for voice or data, and they allow connections like any other phone line. The D channel is used for signaling and also allows for low-speed, packet-data connections (these uses of the D channel are explained in Chapter 3). The B channels are capable of transmitting at 56 or 64 kilobits per second (kbps), and the D channel can transmit at 16 kbps, as illustrated in Figure 1.1. (Unlike Basic Rate ISDN, Primary Rate ISDN uses multiple B channels to provide faster connections.) Compare this with analog fax and data transfer at modem speeds of 9.6 kbps with only one channel.

FIGURE 1.1:

Basic Rate ISDN provides three communication channels on the same line.

What this means to you is that you will have faster connection times and also more reliable transmissions. It will take you much less time to set up your call to transmit information or data and send it to your office or remote location. It also means that the repeated retries that are usual events with your current telecommunications service should be significantly reduced, if not completely eliminated.

With its three channels in a single line, ISDN allows multiple transactions or conversations at the same time. This means that you can, for example, check e-mail, call someone to discuss the business for the day, and fax a letter or registration form to someone else at the same time (with the proper PC software and hardware for your ISDN service, of course).

We will go into more detail about the advantages of ISDN in Chapter 2 and its technical description in Chapter 3. But it is important to keep the two main characteristics of ISDN in mind:

- It is a fast and reliable method of sending and receiving information.
- It allows you to multitask at high speeds.

These two dimensions of the service are what make ISDN a more efficient method than current analog service, and they also allow you to be more productive.

A Brief History of ISDN

Since the early days of the development of this digital dial-tone service called ISDN, there have been some major obstacles to its acceptance by the telecommunications and computer industries. It has been somewhat of a technology looking for an application.

ISDN began in the late 1960s with the work of the CCITT (Consultative Committee for International Telegraphy and Telephony) on the IDN (Integrated Digital Network), an effort to standardize the existing digital telephony networks. The IDN was to be only an internally digital network; that is, the local loop—the connection between a residence or business and the local phone company—remained analog. At the time, the IDN was in the hands of SSG-D (Special Study Group D), a method used by CCITT until the members were sure that creation of a full study group was warranted.

NOTE The CCITT is an international organization that is responsible for many standards used in communications, telecommunications, and networking. For example, this organization developed the V.42 and V.42bis standards for modems. The CCITT is part of the ITU (International Telecommunications Union), a United Nations body. Recently, this committee has been renamed the ITU-TSS (International Telecommunications Union-Telecommunications Standardization Sector).

The "new" beginning of ISDN can be attributed to a paper submitted to SSG-D in 1972 by a Japanese group. This paper proposed that IDN be the single network that would cover all services, since any information can be represented in digital format (and the paper's authors used the abbreviation ISDN for the first time). This plan required making the local loop digital as well, and so ISDN was born. SSG-D also attained full status as Study Group XVIII and was recently changed to Study Group XIII.

The Breakup of AT&T

During the early 1980s, the CCITT conducted the early research and development that lead to the service that was the beginning of ISDN. AT&T (Bell Labs) was one of the first companies to implement the ISDN standard. During this time, the new protocol that allowed multiple channels of communication on a pair of copper wires started to develop into a technology and standard that showed promise in the commercial markets.

It wasn't until 1984 that the phone companies started to take on this new technology as a possible new service offering. It was also during this time that, as part of the Justice Department's modified final judgment (MFJ), AT&T was being broken up into seven regional holding companies, the Regional Bell Operating Companies (RBOCs). These companies had been part of the Bell System phone company, referred to as Ma Bell, for the previous hundred years.

The breakup of AT&T could be considered a contributing factor to the slow development, rollout, and marketing of ISDN. All seven RBOCs needed to individually develop, test, and package the technology prior to making ISDN commercially available. Before the breakup, AT&T had done all the major product development and passed down the product plans for the local Bell companies to simply execute under the direction of AT&T headquarters. The RBOCs did not have much background or experience in product development and marketing prior to 1984, so they had much to learn.

Also, as part of the divestiture's restrictions, the RBOCs could not manufacture telecommunications equipment or customer premise equipment (CPE)—something AT&T had done historically for all its services. By designing and manufacturing customer equipment, AT&T had been able to make sure that equipment was compatible and also available at the same time as a particular new service was being introduced. Most customers could be sure that the first phone service was compatible with their first phone. They could also be sure that it was available only in black.

NOTE This brief history of ISDN is written from the perspective of someone who understands the RBOC and Bellcore technology and product development picture. We have been a part of these early years of technology trials and tribulations. Bellcore is the research and development consortium formed by the seven RBOCs as part of the 1984 AT&T divestiture. It was meant to fill the gap created by AT&T Bell Labs, which stayed with AT&T. And now AT&T is splitting again into three different companies: Carrier (AT&T), Computer (NCR), and Network Equipment (Lucent Technologies).

Early Development

During the 1984 to 1986 period, the various RBOCs were experimenting with how to test, trial run, and develop new services on their own. ISDN was one of those first new technologies that was tested and developed during this period of experimentation. These phone companies were trying to understand marketing and how to judge customer demand. They were learning how to advertise, introduce, and provide the entire solution, with much of their attention focused on analog voice applications.

ISDN for Small Phone Systems

The early development stage of the ISDN technology went in many different directions. There was a time that ISDN was looked upon as a great way to increase the use of small phone systems, such as key systems and PBXs. The thought was that the RBOCs could convert customers of these services to ISDN digital systems and also increase throughput and features of the lines. It looked like an easy way to satisfy the needs of the business customers who wanted small phone systems with features like call-forwarding and digital connections.

Many companies saw that Primary Rate ISDN provides digital, full-duplex (two-way communications on the same channels) connections. This could be used for an intelligent trunk line, which could efficiently and dynamically route calls between PBXs or between large business sites and the long-distance carriers.

> **NOTE**
>
> Primary Rate ISDN (PRI) consists of an aggregation of multiple B channels. In North America and Japan, the most common PRI service uses 23 B channels and 1 D channel (64 kbps), which is the equivalent of a DS1 service in the United States. In Europe, the most common PRI is 30 B channels and 1 D channel, which is equivalent to the E1 service in those countries. Chapter 3 provides more details about the different types of ISDN and communication channels.

Primary Rate ISDN also could eliminate the number of interface cards for PBX trunk connections. Instead of multiple DS0 (single line) channels, you could have one Primary Rate ISDN line with one interface card and 24 DS0 channels (30 DS0s in Europe). This saved a lot of hardware cost for the PBX and was technically more functional.

COs, Switches, PBXs, Centrex: Some Definitions

The phone companies have locations in each community called central offices (CO). The CO is where the copper wire goes from your house to be connected to other phone subscribers and long-distance carrier networks. These COs house a switch to connect one user to another user or to another network. This switch is really a large, complex computer that manages all the connections and provides features like call-waiting, three-way calling, and so on. These switches are the equipment that needs to be upgraded to provide ISDN service. The good news is that most of these switches are already digital. During the last five years, all phone companies have replaced older analog signaling switches with the newer digital signaling switches.

PBXs are phone systems for businesses that typically allow multiple lines to act as one phone system. Each line has similar voice features (such as call-transfer and call-forwarding). Most businesses prefer having a system that ties their lines into one system like this so that they can minimize the phone connections, or trunks, to the outside world. It also gives businesses a flexible phone service, with each person getting a dedicated line and features. *Key system* is a dated term for a small PBX that has two to five phone lines, with lighted buttons on the phone to show which lines are in use.

Trunk lines are the lines that can carry multiple phone lines or channels to the phone company's CO. Typically, in a large PBX system, there are 7 phone lines sharing one channel, and each trunk line can carry

24 phone lines or channels to the CO. For older PBX systems, you needed to order trunk lines for incoming calls and other trunk lines for outgoing calls. With Primary Rate ISDN, you can use the same trunk line to dynamically manage incoming and outgoing calls on the same line.

Centrex is a business phone service that provides multiple features, such as call-waiting, speed-calling, and call-transfer, to a group of phone lines for a business. It usually costs more than regular phone service to install and requires a minimum of 10 lines.

ISDN and CO-LANs

One of the other distractions from developing and pushing ISDN during this period was the focus on CO-LAN (central office–local-area network) technology. Many RBOCs were experimenting with some LAN and DOV (data-over-voice) technology at this same time. The early CO-LAN technology testing and market trials took some of the attention away from ISDN development. It looked like the CO-LAN product could provide the remote access for office and home locations. What really happened was the explosion of the LAN market in businesses. Customers put the LAN on their own site and did not buy the central office products that the RBOCs were offering.

Although ISDN could be configured to be somewhat of a LAN substitute, it would not work that way when the users were in the same building. For those single-site LANs, there was no need to send the wire in and out of the building to the central office.

It became clear that, rather than have ISDN compete with a CO-LAN product or try to create an ISDN LAN product, ISDN should be developed as a way to remotely connect users to their existing office LANs and to connect LANs to each other. This interconnection of LANs was part of the growing internetworking explosion during the late 1980s.

ISDN as a Mini-Centrex or "Enhanced" Centrex

During the late 1980s, ISDN voice applications had some success, and there was even some work on developing ISDN to become the new and improved Centrex. The voice features and digital character- istics of ISDN seemed to make it a great way to compete against the PBX threat.

Along with the idea that ISDN could replace Centrex was that it also could be used to create a "mini" Centrex, or a small business phone system for companies, as we mentioned earlier. It was during this time that using ISDN as a virtual key phone system had some popularity. Although this application had some success for awhile, it was too com- plex for most small businesses, and some central office switches did not support it in the high-volume traffic of business environments. There is still some work going on in this voice-application area (for example, Siemens—a German-based telecommunications manufac- turer of voice, data, and telephony equipment—recently announced new products that use ISDN as the basic networking technology), but most do not see it as the biggest market for ISDN services.

NOTE PBX use will be a niche application for ISDN, but it will never approach the data applications that we describe in this book. It also will always be limited to only certain segments of the business community, which will have many other alternatives for small phone systems.

Although these various efforts show there was some confusion about the development of the ISDN service, there were two key beneficial results:

- It was proven that the service has many features and functions that allow multiple uses of the technology. For example, its multiple channels can be assigned different phone features and numbers.

- Developers learned that basic general availability and ease of use will increase the demand for any network service.

ISDN Standards Development

ISDN has gone through many standardization efforts and versions, which have contributed to the delay in making the service more available and easy to implement. The ISDN standards were not static during the early period of ISDN development. Some of the tests and experimental uses led to new revisions or modifications to the service itself. These early tests involved the physical characteristics of the ISDN bandwidth; that is, the developers simply tested having two B channels and one D channel on the same copper pair of wires to confirm that the ISDN protocol actually could be implemented and provide simultaneous, reliable digital service.

During the late 1980s, there were multiple versions of ISDN being tested and used on a trial basis. These early versions each had some incremental features and configurations that were being introduced. Some were voice features for increased phone functionality, such as priority ringing and call-forwarding. Many features were proposed by the phone companies and by the customers trying out the service. Some, like the common set of voice features, were quickly developed and implemented. Other features, such as multiple terminal identifiers and multipoint (which allow multiple terminals to share an ISDN line for data applications) are still being refined. Meanwhile, the companies that provided the ISDN switches for the phone companies were trying to develop the ISDN service with the early standards that were available at the time.

In Chapter 3, we will review the standards that apply to ISDN and the development of one National ISDN-1 standard. This standard, the Bellcore National ISDN-1 specification (SR-NWT-001953), has now been adopted by the Corporation for Open Systems, ISDN Executive Council, as the service on which to base the switch manufacturers' ISDN service.

Early ISDN Availability

As we mentioned earlier in the chapter, in the late 1980s, development of ISDN was geared toward the business service, Centrex, which was one of the major business products for the phone companies at the time. The switch vendors developed ISDN as an additional feature or option to this business service. This made the Centrex service a requirement for ISDN service.

Many customers who wanted to buy or try out ISDN were forced to purchase Centrex to get ISDN service. The result was that the addressable market for ISDN was business Centrex customers, typically with many phone lines. This led to the early testing and applications of ISDN focusing on large business users; very little was developed with the middle-size or small business market in mind. Because its availability was low and its cost was high, few customers were interested in using the service.

The early pricing and tariffing of the ISDN service were complex and differed greatly by area and RBOC. The first RBOCs or phone companies to tariff ISDN service, during the late 1980s, were Ameritech, Pacific Bell, and Bell Atlantic. Many of the other phone companies, including independent companies, followed up with their own ISDN service options. The various phone companies' service offerings were very different in features and prices. This made it extremely difficult for customers to understand the features and cost of the service. Only some areas had ISDN available, and it was only as a business service, with no options for residential service.

ISDN Hardware Development

The telecommunications hardware and computer industry had little interest or commitment in the ISDN service in the beginning. The initial customers were offered few hardware solutions with some clumsy software solutions provided by a very few vendors. In fact, many ISDN customers had equipment that was made by the limited number of ISDN switch manufacturers.

Most of the early ISDN applications were centered on voice communications. Few of the early users were concerned with the data-transfer capabilities of ISDN. Therefore, there was minimal support from equipment vendors other than voice phone set manufacturers. Very few of these units had any data-application features.

Another obstacle to ISDN equipment development stemmed from the AT&T breakup. As explained earlier, as part of the divestiture regulations, the RBOCs could not manufacture customer equipment. This meant they could not design the hardware to be compatible with the features and functions of the ISDN services the companies offered.

In the early implementations, the hardware required to connect to ISDN was expensive, complex to install, and composed of multiple hardware units. It took some time to modularize and miniaturize these components to fit into one piece of hardware.

The early pioneers of ISDN started with the networking terminating device (called NT1) and the power supply, and then added the ISDN terminal adapter. Can you imagine having a modem that came in three parts for you to install and configure on your PC? That is what the early users of ISDN had to endure to get the service at this time. Today, the NTI, power supply, terminal adapter, and analog converter are in one integrated unit, or you can simply use a PCMCIA card for the connection. Figure 1.2 illustrates the difference between the early ISDN hardware configuration and the current equipment.

FIGURE 1.2:

ISDN hardware has evolved from a complex configuration of multiple parts to a single integrated unit.

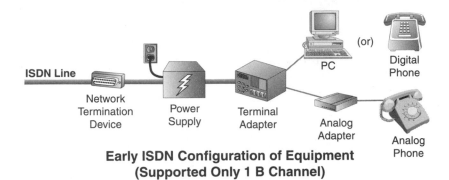

**Early ISDN Configuration of Equipment
(Supported Only 1 B Channel)**

**Current Equipment Configuration
(Supports 2 B and D Channels)**

Comparing Modem and ISDN Equipment Development

The evolution of modem technology and the development of ISDN technology have quite a few similarities. The early modems had separate power supplies, and the main modem unit had to be kept at some distance from the computer. These units were expensive and cumbersome.

Consider how long it took for this analog (not digital) transformation of remote access for the PC. How long did we live off the modem-pool arrangement because modems were too expensive for everyone to have a dedicated modem in the office? It took well over seven years for the modem technology to become common and reliable enough for most users to feel comfortable dialing out to anywhere and having access from their office or home.

The modems in PCs are now so integrated that you do not need to even think about the modem settings or configuration. The multitude of modems have gotten so "Hayes-compatible" that nobody has a Hayes modem anymore. The modem manufacturers have realized that modems that are truly "plug and play" are the ones that are selling. PC manufacturers know that the only way to introduce network connectivity in a PC with a modem or any network interface is with equipment so economical and so integrated that most users can start connecting to remote computers and resources as soon as the PC is out of the box.

The acceptance of modems is similar to the movement toward ISDN during the past few years. However, there are some differences that have made it more of a journey to the small business and mass market than it was for modems and analog dial technology.

First, analog dial-tone as we know it today is universally available (for the most part), and it has been available to most people on a dedicated basis for the past 20 years or more. We assume that everyone has some phone service accessible to them.

The second major difference is the multichannel capability of ISDN versus the single channel of your regular analog service. Although there are technologies that can allow multiple channels on the same analog connection, most people use the dial line for one conversation or one connection at a time. With ISDN, the concept of multiple transactions or conversations at the same time introduces complexity and some changes that consumers can't easily understand or use. How does it make multiple calls at the same time? Why would I want to do this? Isn't it cheaper just to have two lines? Availability and understanding of the service being offered are two of the major differences in the development of modems and ISDN.

Early Marketing Approaches

During the 1980s, there was a lot of marketing hype of the "killer app" (mass market application) for ISDN. It went from things like residence telemetry of your home power usage and alarm security, to electronic phone sets that were able to provide more features to telecommuting workers at home. Preprint press industries and any business that transferred large data files were among the many that were the target of ISDN marketing devotees. The shortcomings of the availability of the service, the complexity of the installation, and the cost were not considered in these promotions. Some customers made a large commitment to the technology, only to discover that it fell short of their expectations.

Many of the initial customers were large companies who were interested in the new voice features and phone sets that came with the early implementation of ISDN. These offerings attracted many customers into individual custom contracts. All RBOCs started with large business customer contracts to introduce the new ISDN service.

Even AT&T Network Systems created some early sales by having all of the company's regional offices order ISDN from all the RBOCs across the country. This did generate some business, but its effects were limited. It mainly lead to major deployment in just the areas surrounding the locations of those large customers. This promotion of ISDN did show AT&T's commitment to the technology, but it was somewhat tainted by the fact that AT&T had the most to gain by its success.

The transition cost was still very high at that time, but a few users and companies felt it was worth the investment. Many of these early users were also provided some special offers of service and/or equipment to make the change more attractive.

As mentioned earlier, there were also developments of the ISDN as LAN service as a competing technology. The early LAN directions did provide great experience and input for the remote access and

LAN interconnection capabilities of the service. And now remote access and LAN-to-LAN communications are the growing applications for ISDN.

Another important aspect of any new communications service is that it needs to be backed up by support and service resources. New users need to be able to contact trained personnel for help. Customers should get all the information they need for smooth implementation and troubleshooting. This support was severely lacking for the communications service and the equipment during the introduction of ISDN.

In summary, the early efforts to market ISDN were marred by the multiple implementations, low availability, and limited acceptance by the hardware manufacturers, computer companies, and online service providers. Even when one standard became prevalent, it still had multiple implementations. If you had to identify a recipe for a new service that would not be attractive or fast growing, the early development days of ISDN would be it. This technological melodrama is an abbreviated journey through ISDN development during the late 1980s.

The 90s: The Situation Improves

It was not until the 1990s that things started to change in the way that carriers and RBOCs were offering ISDN service. At that time, the ISDN technology started to get increased attention from the computer and network hardware industries.

During this period, many of the customer requirements for service and support started to fall into place. The carriers and RBOCs also began to see that there were many horizontal applications that could appeal to many markets and started to focus on a large set of these markets. They began to improve the critical factors of cost of deployment, ease of implementation, and overall availability. This was not just availability of the ISDN service in major population centers, but also availability of the equipment that customers needed for service and availability of support centers that could explain the ISDN service, equipment requirements, and configuration of that equipment.

In November 1992, a national event called TRIP '92 (TRanscontinental ISDN Project) was the first demonstration of the use of ISDN applications across the country by major carriers and equipment providers. This event was based on the National ISDN-1 standard and created the momentum for implementing that standard as one universal interface. The demonstration showed the carriers that there were good reasons to expand the service to cover more than just small metropolitan areas in each major city.

This is the about the same time that "extended" ISDN was being tested and deployed. New technology allowed for extending the distance from the central office for ISDN service. Prior to this, ISDN service was limited to a physical distance of 15,000 to 18,000 feet from the phone company's central office. AT&T and Adtran are the two major vendors who have developed and delivered equipment that allow the carriers to extend the ISDN service.

Another major development that increased the availability of ISDN was the redesign of the product so that it could be provided on a single-line basis. No longer was it something that required Centrex service, as it had been in the past. During 1994, a few companies, starting with Pacific Bell, began to offer residential ISDN service. By the end of 1996, all the RBOCs and local carriers will offer some form of residential ISDN service.

In addition, you will see offerings of ISDN from the competitive access providers (CAPs) in not only major metropolitan areas, but also in some smaller cities across the country. Many state regulatory commissions are seeing the need to treat ISDN as the initial common access (or "on ramp") to "information highways." For this reason, they are asking the local carriers and the competitive network providers to make ISDN available to as many customers as possible.

The terminal adapter equipment manufacturers also started to work toward many of the data applications for ISDN. They looked at new ways to use ISDN for LAN interconnection, disaster recovery for data, and desktop videoconferencing.

ISDN has now become an important product for the phone companies and competitive access providers, as well as an attractive access method for many of the new information service providers. It is a common method of accessing online services and remote databases for all types of users and companies.

The Current ISDN Market

So what has changed in the last few years that has made a difference in the acceptance and demand for ISDN? Has it been the marketplace? Has it been the computer or PC industry? Has it been the services demanding remote computing? Has it been the marketing skills of the carriers or RBOCs? The answer is that all of these factors have contributed to the growing demand for ISDN services.

The increased PC ownership in businesses and most households has created a large market for networked computers and desktop remote computing and commuting. The trend towards this "computing and commuting" is driving the need for better ways to connect remote computers and access resources.

Many online service providers have seen rapid growth in their market. These online services—Internet service providers (ISPs), America Online, Prodigy, CompuServe, and so on—are now looking at ISDN as a new digital option to access their services. Most are beyond the study and trial phases of ISDN and are advertising complete sign-up packages, with ISDN hardware and software included.

If you look at the main areas of overall market acceptability, you will find four major shifts in the market that have increased ISDN demand and acceptance:

- Increased ISDN service availability

- Advances in ISDN hardware

- Growth of applications

- Availability of ISDN packages, with bundled hardware and software to provide ISDN services

Computing and Commuting: Where's It Going?

We've seen announcements about a $500 scaled-down PC model that can be used as a video, text, and audio terminal. Oracle is making some major advances in its multimedia projects. The advent of the affordable multimedia machine points out that this PC ownership trend will continue and increase to not only a single PC, but multiple PCs in the home.

Recently, in the Northeast area during a major snowstorm, the local phone companies noticed a new phenomenon. For the duration of the storm, while schools were closed and highways were shut down or slowed to one lane, the phone network took an unexpected spike in usage. This seemed to be the first time that the communications from many remote and home users were generating enough traffic to create a peak in network utilization.

The mobile, home, remote, and part-time work force is continuing to grow, with increases of between 5 and 20 percent annually. This is the SOHO, MOHO, and ROHO (for small office, mobile office, and remote office home office) market you've been hearing about. Whatever acronym you use for this market, it is now common for corporate, contract, and self-employed full- or part-time professionals to work at home and remote sites.

Online services have experienced explosive growth. The number of subscribers to America Online has doubled and has recently exceeded 6 million. The number of Microsoft Network subscribers quickly hit 500,000 users and is now on its way well past the million-subscriber mark.

Quicken (by Intuit) has started a financial network service, which is now a standard option of this money-management software. This is just one more service that many of us can access from our homes.

These are just a few examples of the trends that are moving more people to need a higher-speed access from their homes to remote locations, and to a larger variety of remote locations. This is why the "any-to-any" connectivity of ISDN is so attractive to most users. In most cases, you can use ISDN for many remote services, not just to connect to one dedicated or preconfigured location. In other words, you do not need to be limited to going to only the Internet or just to Prodigy.

Market Shift 1: Service Availability

ISDN service is now available in all the major cities in the country. It is available in about 90 percent of all the metropolitan areas and more than 75 percent of the suburban areas in the United States. Of course, this coverage varies by region of the country and by local phone company, but its availability will continue to increase for at least the remainder to this decade.

This means that ISDN is available to the majority of end users who need the remote connectivity. There are many studies that talk about where the remote users live, but it is safe to say that at least one-half of this market is within the areas where ISDN is available. And the service is expanding as each carrier continues to deploy and invest in digital-switching technology. Also, since ISDN is a standard protocol, it actually can be transmitted over various media, so satellite- and fiber-transmitted ISDN are feasible.

This is much different from the early days, when very few areas had ISDN available. Even as of two years ago, many phone companies did not offer the service in some major cities. There were also some that did not do any kind of advertising of ISDN in their areas. Now all seven RBOCs across the country, along with independent companies (such as GTE, Centel, and so on), have the switching equipment to provide ISDN service.

In addition, there have been provisioning improvements in the ISDN deployment itself. In the early days of ISDN service, the local central office had to be ISDN-capable for you to get service. Most local carriers now provide "extended" or "foreign exchange" ISDN, which means that your local switch may not have ISDN available but it can be brought into your area by extending service from an adjacent central office. This has also been called "ISDN anywhere" service, although the term is deceiving, since ISDN still has some technical limitations (but a name like "ISDN almost anywhere unless you have too many technical and physical distance limitations" service is not likely to catch on).

Another improvement in this area is the introduction of a card that extends the distance of ISDN service availability. Previously, if you were beyond 15,000 to 18,000 feet away from your phone company's central office, you could not have ISDN service. This limitation has been overcome with the use of BRITE (Basic Rate ISDN Terminal Extension) cards and ISDN extenders in the copper loop going to your house or business. These developments allow even more homes and more distant locations to have access to ISDN service. As mentioned earlier, AT&T and Adtran have been leaders in this extended ISDN capability.

You'll read more about the growing availability of ISDN in the next chapter. The increase in ISDN service availability leads to the second area of market acceptance for ISDN.

Market Shift 2:
ISDN Hardware Advances

Equipment manufacturers, network hardware companies, and the computer industry in general are embracing ISDN as the next new communications method. Many of these companies were not so sure about ISDN during the past few years, because they were getting weak or mixed messages, at best, from the RBOCs and service providers. The service providers were waiting for more demand to deploy, the terminal equipment manufacturers were waiting for more deployment to develop, and the early users were waiting for both to increase demand. This is a typical stage in a communication service product cycle.

These companies were on the sidelines, but within the last year, they have decided to take the leap. Many vendors are now producing new hardware for the ISDN service. Not only do you have more ISDN hardware choices, but you also have access to more technical support and service for the proper installation and use of ISDN.

Many of these companies have bought or are now buying the technology from other companies (or simply buying the companies) to speed the manufacture of ISDN equipment for many of their network interfaces. Witness Cisco Systems purchasing Combinet, 3Com purchasing Access Works, Bay Networks purchasing ISDN Systems Corporation, and Cabletron purchasing Network Express.

Many companies are building their entire future on ISDN hardware and service. For example, Ascend Communications started as one of the first ISDN inverse-multiplexing vendors for room-system videoconferencing equipment. It has now quickly expanded into the ISDN remote-access and Internet-access market. Companies are betting their high-tech farm that this ISDN growth will continue and provide enough of a market to produce high profits for many years to come.

Most, if not all, modem manufacturers now have ISDN technology to either complement or supplement their modem product lines. They have seen that ISDN is one of the next digital services that needs to be integrated into communications systems. Motorola and US Robotics have invested in the fast modem technology but also realize that the next improved access method is ISDN.

Many hardware or PC manufacturers will include ISDN in their next round of upgrades for the PC and network communications alternatives. There is already work on a PC card for laptops (which Intel, PictureTel, and 3Com have announced). This will create additional demand to fuel ISDN service availability and reduce overall service cost. Most, if not all, network LAN servers will come with ISDN as a primary or backup connection for interconnecting LANs or communications dial-up connection.

It is very probable that ISDN will become a standard interface in addition to the analog dial connections today. We will see the modem and ISDN technology combined. This will make it much easier for the user to access remote resources and not worry about whether the network connection is working in the analog modem or digital ISDN mode. As the industry players in the PC and the internetworking worlds continue to buy into ISDN technology for their products, it will be integrated into many everyday services, equipment, and business applications.

Market Shift 3: More Applications for ISDN

How many commercials have you seen for online service providers such as America Online, CompuServe, and Prodigy?

How many flyers have you received for online banking and stock market services accessed through your home computer?

How many ways do you have to electronically access information immediately through your home computer for work or play?

NOTE

Microsoft has escalated this end-user race with Microsoft Network (MSN), one of the main new features of Windows 95. It can also be seen as the next Microsoft initiative to wire Microsoft users together for product information and service usage statistics. What better way to track software and software usage than to have most of your users connected by an online service? What better real-time marketing intelligence than to have users connected for your surveys and market research?

It seems that every man, woman, and child on the planet is now aware of the Internet. The media has covered the Internet as the next information wave that everyone must ride to stay informed in today's society. Schools are being wired for Internet access and communication so that our kids will be trained to communicate worldwide from their PC and perform research without leaving their desk at home. It will not be too long before the PC will be used for more curriculum and more communication among the members of the educational community—parents, students, and the community at large.

There has also been an enormous growth in the desktop videoconferencing market during the last two years. Along with the video capability, desktop conferencing has ignited the data collaboration applications. These applications allow remote users to work together on projects from afar. They can draft, edit, and produce documents jointly, without being restricted by physical limitations or geography. For example, I can share this book and its charts with editors across the country and not be concerned about when they can come to one coast or the other. Project teams and other collaborative work efforts can increase their productivity while decreasing work time.

Another of the application areas that has grown is the remote access to LANs. We mentioned earlier that the growth of LANs in the 1980s contributed to the confusion of ISDN being developed as a LAN substitute. The recent demand for business and home remote access to corporate LANs is as strong as the demand for the interconnection of LANs. Both of these applications have increased the need for the ISDN service. The switched characteristic of ISDN provides an excellent bandwidth-on-demand capability for remote users and LAN administrators, connecting users to LANs and LANs to LANs.

NOTE You will read more about bandwidth on demand in Chapter 3. It simply means that when you or your LAN system needs a connection, the system can dial up the remote location when it sees the demand for the connection. One of the most common applications is your e-mail system. Most LAN e-mail systems are programmed to look for incoming and outgoing mail every 15 minutes. When the system sees an e-mail message for a user at a remote site, it can dial up that site and transmit the e-mail to that user's system.

These online services, desktop computing, and remote-access applications need some capabilities that ISDN can provide: shorter connection time, increased transmission speed, and reliability. This is what makes ISDN a very attractive communications method for the business community and mass markets. These specific applications are covered in detail in Part 2 of this book.

Market Shift 4: The Packaging of ISDN

The fourth major factor to make ISDN service attractive to small business and consumer markets is packaging. This starts with limited availability and trials of the service with equipment to provide complete application solutions. For example, you'll see ISDN equipment (like the Motorola BitSURFR and BitSURFR Pro) in Egghead Software stores. This represents the beginning of the complete commercialization of ISDN service.

In addition to the computer industry and retail channels, many of the RBOCs and carriers are promoting ISDN with packages and service support to simplify the installation process. They have added many phone support centers and field support resources to provide the assistance needed to smooth the transition from analog to digital network connections.

The computer software and hardware manufacturers have been working together to provide an integrated solution for the general business and home user. For the mass market to genuinely accept any new technology, it needs to be easy to purchase, install, and maintain. In other words, the technology must be "packaged" for easy consumption for the business or home user.

Consider how the industry has packaged modem communications. When was the last time you went shopping for a modem? Did you ask about speed, number of transmission errors, or guaranteed throughput? Did you ask if it automatically adjusted its speed based on the transmission degradation? Did you ask the make and model of your modem to be sure that it was compatible with the various services that you were going to use it for?

If you are an average user, it is highly likely that you have never shopped for a modem, and you probably don't know anyone who has. You assume that the modem that came with your PC can handle most, if not all, dial-up connections that you will encounter. The details of modem technology are handled by computer network communication specialists, not end users. This is how it will be with ISDN as desktop PCs and other network communications equipment integrate this interface.

Most major router vendors now offer ISDN as a standard primary and backup network communications option. It is fully integrated into the router configuration and backup contingency software options. All videoconferencing units now come with an ISDN interface as a standard interface. All dial-up video room systems have ISDN inverse multiplexers for connections with anyone across the country and the world.

It is unlikely that you will ever shop for an ISDN terminal adapter just to try out ISDN service. As we've already said, the next PC you purchase will most likely have an ISDN terminal adapter with modem functionality built into the same unit. You won't know which network connection method is being used with your particular service. And if it is packaged correctly, you won't care. You will only notice that your network connection seems to be faster. You will see that surfing the Net will be much quicker as you move through multiple Web pages to find that vacation information or ESPN contest form.

CHAPTER

TWO

2

Why ISDN Now?

- The differences between analog (regular telephone) and digital (ISDN) communications

- Extended call features that are available with ISDN

- How ISDN can reduce communications costs

- The speed and quality of ISDN transmissions

- Indicators of ISDN's use in the United States

- Alternative telecommunications options

- When to consider using ISDN

In the previous chapter, you read about some of the events that lead to the recent demand and interest in ISDN service. But you may have some valid questions about this service: When do you need it or even need to consider it? Will this service really be more available and more economical? What signs are there that installing ISDN will be almost as easy as Plug and Play?

We believe that ISDN is finally going to make it into the mainstream because there now are enough applications and ISDN service providers for most people to consider using it. And ISDN now has enough industry-wide support to make it an option that is economically viable.

In this chapter, we'll take a look at the reasons why you should consider using ISDN. We'll begin with the difference between digital and analog communications, and in particular, how ISDN service compares with regular phone service. Then we'll focus on the advantages of ISDN: speed, cost-effectiveness, and high quality. Finally, to be fair, we'll see what other communications options are available.

The Digital Difference

Today, more and more people are buying modems to communicate with the office, surf the information superhighway, communicate with each other, or even play games. Typically, the modems operate at 14,400 or 28,800 bits per second (bps). If text were the only media people shared over modems, 14,400 bps would be really fast. For example, a 14,400 bps modem takes just over five seconds to send a 1000-word article.

But today's communication is more than just text and words. We share pictures (graphics), moving pictures (animation or video), and sounds. Everywhere you look, media is turning digital. Digital cameras capture images for us and our computers. The days of generating our music from a record player with a vibrating needle are gone. A laser light reads ones and zeros reflected from shiny rainbow compact discs. The old adage "A picture is worth a thousand words" comes up short in the digital world. For example, a high-quality graphic image, *one-fourth* the size of a computer screen can comfortably exceed a 10,000-word article. Even five seconds of high-quality sound from a CD or five seconds of full-screen video consume more file-storage space than a thousand words.

Currently, telecommunication experts believe that the analog 33,600 bps modem is probably the upper limit for regular phone lines. If today's pictures are worth 10,000-plus words, we will need to begin looking for new ways to exchange more data, faster. The phone companies refer to our regular analog telephone service as POTS (Plain Old Telephone Service). As you learn more about ISDN, you will begin to appreciate why our current phone system is plain and old. One way to start appreciating the differences is to compare regular telephone with digital telephone.

Regular Telephone versus Digital Service

The benefits you can enjoy by switching to ISDN service depend on why you are considering the service and how you intend to use it. In the following sections, we will compare traditional telephone service with digital by taking a look at some common questions that occur to most people when they are first introduced to ISDN:

- What, no dial tone?

- What do you mean by two lines in one, and for voice *and* data?

- What are extended call features?

What, No Dial Tone?

Did you ever wonder why you can call the power company in the dark to tell the customer service representative that you're without power? The phone company provides power to your phone set to maintain the dial tone. Long ago, the dial tone was needed to indicate if a line was functioning. Today, a dial tone remains a familiar audible indicator that assures us humans that our phone is working. The missing dial tone is a often the first major difference between regular telephone (analog) and ISDN (digital) that customers notice. Although it's unnecessary today, most ISDN equipment still generates the familiar dial tone for us when an analog device is attached.

Because ISDN equipment consumes more power than POTS equipment, power delivered through the telephone network, to generate a dial tone, is not feasible. ISDN equipment needs an external power outlet just like your television, stereo, or computer. As an ISDN user, if your residence or business experiences a power outage, your ISDN equipment will be without power, and you won't have your digital telephone service.

WARNING ISDN needs an external AC power outlet, which means that you will be without ISDN service during a power outage. Therefore, you may want to consider keeping your analog line as a backup.

Two Lines in One—
For Voice and Data?

Families with teenagers and home offices will appreciate the ISDN benefit of getting two lines in one. With ISDN, the same twisted-pair wire that has traditionally carried only one conversation in an analog environment now carries two separate voice conversations at the same.

With two separate B channels, you can conduct a conversation on one channel and use the other channel for another voice call or a data transmission, such as from a personal computer or fax machine. The additional D channel can carry data at the same time.

You can use the same line to transmit and receive voice, data, images, and video simultaneously. With a traditional analog line, you need to connect to each of the services individually.

What Are Extended Call Features?

Depending on the ISDN equipment you purchase and local availability, ISDN provides extended call features (such as call forwarding), which you can access by pressing a button or by dialing an access code. If you have a regular analog line, you need to pay extra to receive the same call features.

Customers who wish to take advantage of extended call features will discover a wide range of call-management features. For example, you can put multiple calls on hold, transfer calls to other numbers, create conference calls, forward all calls to a number you specify, display the number of the person calling (even if you're on the phone), block calls from a specified number, or automatically call back the last person who called. The following is a sampling of the range of call-management features available (the number and type of features will vary by region and availability):

- **Call Forwarding:** Forwards calls to a preselected number when the called number is busy, or after a preset number of rings.

- **Call Pickup:** Lets an incoming call be picked up at another phone, answering machine, or station.

- **Directed Call Pickup:** Lets calls from a specific extension be automatically forwarded to a second number.

- **Message–Waiting Indicator:** Shows that an incoming voice message has been received.

- **Directed Dial:** Incoming calls can be automatically forwarded to a central office, a car or mobile phone, or a phone at another location.

- **Ringing Options or Distinctive Ringing:** These options include abbreviated ringing, delayed ringing, no ringing, and normal ringing standard.

- **Additional Call Offering:** Allows the company's telephone switch to use a distinctive ring to signal that a call wants to come in, even if both B channels are in use. The caller can decide to terminate a call or put one on hold to open up a B channel for the incoming call.

The extended call features in ISDN allow you to "push" functions back "down the wire" to the phone company's central switch. The call features normally performed by a company's PBX are now performed at the phone company's central switch. This is a very important feature of ISDN. ISDN's signaling capability is always alive, available to do your "bidding." Today's modern digital switch is a powerful device that is capable of many service features (some of the more common are listed above). ISDN equipment can offer a wide range of voice features and benefits without requiring you to install expensive switching equipment.

Advantages of ISDN

In the following sections, we will examine several ways that ISDN can make your communications faster and more productive. We'll also see how the quality of ISDN transmissions compare with analog transmission quality and how using ISDN can be cost-effective.

Digital Speed versus Analog Speed

ISDN is fast! Digital signaling increases speed dramatically over analog signaling. For example, no conversion is needed to convert a digital signal for use inside a digital network. Figure 2.1 contrasts the speed of two analog modems with ISDN equipment.

FIGURE 2.1:

Compare the speed of analog and digital equipment.

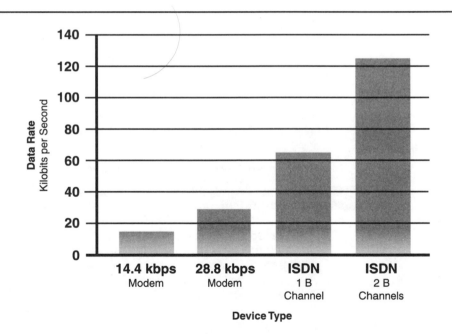

Before ISDN, normal phone lines carried only 2.4 kilobits per second (kbps), reliably. This speed is approximately one-third of a single-spaced, one-page document, or one page every three seconds. After ISDN, digital phone lines can carry 128 kbps over the same wire, or 22 pages every second! (Each B channel carries 64 kbps, for a subtotal of 128 kbps; the D channel carries another 16 kbps.)

WARNING
Unfortunately, you cannot "combine" two B channels (64 kbps each) and add a D channel (16 kbps) to get 144 kbps. Some technicians and reviewers in computer publications may portray ISDN as providing a 144 kbps throughput; however, only the two B channels can be combined to give you 128 kbps (64 kbps + 64 kbps). Still, this is four times faster than a 28.8 kbps modem!

ISDN's speed allows very quick file transfers. For example, using both channels, a one megabyte uncompressed file can be transmitted in less than one minute. A comparison of file-transfer times is shown in Figure 2.2. This speed can really make a difference when you are being charged for the online time to transfer complex images, full-motion color video, or other large files. We'll talk more about the cost advantages of ISDN a little later in the chapter.

NOTE
Most ISDN equipment offers 2:1 compression, resulting in an effective transmission rate of 256 kbps. Larger compression ratios are also available. However, your transmission speed may vary, depending on the type of files or data you are sending or receiving. For example, files downloaded from the World Wide Web are already compressed using a compression scheme, such as in uuencoded zip files, JPEG graphic files, or MPEG video files.

In addition to raw speed, ISDN calls set up (connect) much faster than analog phone calls. Analog modems must negotiate the bandwidth (speed) that the phone line will support. In most cases, negotiation can take between 10 to 30 seconds or more! Because ISDN is fully digital, the lengthy handshaking process of analog modems (those screeching, screaming noises) is gone. ISDN calls connect in a matter of seconds.

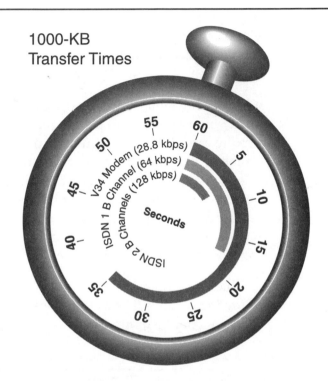

1000-KB Transfer Times

Cost-Effectiveness of ISDN

As electronic communications expand across geographical boundaries to meet the demands of an increasingly global economy, many of us will need to exchange information among a group of widely dispersed users. Whether these remotely located individuals are customers, consultants, or business partners, communication is no longer a luxury. Communication is an essential element in the exchange of information, services, and ideas that give today's business a competitive advantage.

The requirement to communicate with individuals at diverse locations extends beyond a simple voice exchange. Today, we also need to share a wide variety of information in various formats. For example, some individuals may require dialed voice and data connectivity, high-speed data, and file transfer; others may need multimedia and videoconferencing collaborations; and still others will want easy access to centrally archived data and images or even video and sound files.

ISDN's flexibility saves you money by letting you pay only for what you use and avoid paying for leased lines that may seldom be used to capacity. You can quickly reconnect or disconnect from an online service and avoid paying for online charges while the line is idle. Also, your company can improve its communications capabilities while reducing the number of lines it needs; the same lines can handle different types of calls—data, voice, and video—either permanently or as the need arises. ISDN also makes it easy and economical to reconfigure backup and emergency networks.

Table 2.1 summarizes the many cost advantages of ISDN, and the following sections take a closer look at several ways ISDN can improve your communications and save you money, too.

TABLE 2.1:   A Summary of ISDN Cost Advantages

Communications Category	ISDN Benefits
Data	Reduced retransmissions for time and money savings over analog dial-up. Increased transmission speeds at the same per-minute rates. Lower costs than dedicated circuits for low-volume traffic locations.
Voice	Support for call-management features normally found in a PBX. One line that provides two separate voice channels over a single twisted-pair cable. A bonus of packet-switched data capabilities on the B or D channel.

TABLE 2.1:   A Summary of ISDN Cost Advantages (continued)

Communications Category	ISDN Benefits
Network	Simplified network management and maintenance using the public-switched network for voice, data, and packet-switching. Integrated access to voice and data services. Bandwidth-on-demand (bandwidth is dynamically allocated in real time). Reduced costs for international and nation-wide communication. Expanded availability to support digital data applications.
Businesses and telecommunications providers	Reduced infrastructure and maintenance costs by offering multiple services through a single network. Bandwidth-on-demand. Compatibility with European standards.

Communications Cost Savings

For Internet users and telecommuters, ISDN speeds the process of locating data resources and downloading files. A data file that takes four minutes to transmit with the fastest analog modem (33,600 bps) takes less than a minute with ISDN.

Also, ISDN saves money by reducing overall phone and online charges. A typical call setup time of one-half second to three seconds means that users can have almost instant access.

Several ISDN equipment vendors support a configuration that hangs up or disconnects after a line is idle. After a predetermined amount of time that you can define, say two minutes, the ISDN equipment hangs up. Imagine reading a Web page or reviewing e-mail without wasting online time. Ready to move on? A mouse click or selecting a new location signals the ISDN terminal adapter to wake up and reconnect. In an analog environment, using this method means manually disconnecting and then waiting for the slow and frustrating sequence as your modem reconnects. Generally, modems are not very quick when it comes to connecting, or call setup. Waiting for your modem to scream and hiss makes 20 to 30 seconds seem like forever.

NOTE A line is *idle* when nothing is being sent or received through the line. For example, when you stop to admire or read a page on the World Wide Web, your line is idle, similar to a car stopped at a red light. Keep in mind, that new technologies may be sending your computer small files to keep a Web page animation's moving or a text banner scrolling across the bottom of the screen.

Eliminating Expensive, Dedicated Leased Lines

ISDN provides an alternative to the high cost of dedicated leased lines for high-speed digital connections. It offers access to the world-wide telecommunications network through switched digital connections, delivering higher capacity, higher speed, and lower costs than dedicated lines.

The cost-effectiveness is based on lower charges, infrequent use, and the costs incurred only when data is transmitted. By offering multiple services through a single network, a customer can benefit from reduced infrastructure and maintenance costs. For example, because ISDN service is provided through the same copper lines as regular POTS service, most offices and homes are "ISDN-ready." Also, ISDN phones are able to make or receive calls from regular, analog phones.

ISDN's High-Quality Transmission

ISDN is the missing link that turns the last 100 feet between you and your phone company into a digital connection. ISDN transmits data digitally and, as a result, is less vulnerable to static and noise than analog transmissions. Digital transmission means fewer errors, retransmissions (or attempts to send lost data), and less ambiguity (noise, interference, and so on) in your data communications. All of the digital benefits discussed so far—better data quality, higher reliability, and dramatically reduced call setup time—translate into fewer problems and higher speeds.

NOTE Over the past few decades, local and long-distance companies have been quietly converting their antiquated analog systems to digital. The last link between the phone companies and us, known as "the last 100 feet," is what remains of our analog telephone service.

Clear Connections

Data connections are virtually error-free. ISDN is a digital technology, transmitting zeros and ones instead of modem tones over the phone line. As a result, it is very reliable. Analog phone lines (or POTS) pick up noise from their surroundings, which can block connections, slow transmission, and corrupt data. As a result, analog modem connections must dedicate some bandwidth to error correction and retransmission. All this overhead reduces the actual throughput. Without all the fuss (retransmission) and muss (error correction) of an analog connection, an ISDN line can dedicate all its bandwidth to data transmission.

Quieter Conversations

ISDN's better quality is not limited to data transmissions. Digital signaling translates into better voice phone service. Voice conversations over ISDN are crystal clear, having the sound quality of an audio CD. By contrast, analog phone lines, like the vinyl records of yesteryear, are lower quality and noisier than digital.

ISDN's Simultaneous Transmission

Today, POTS permits only one phone call over a single line. Basic Rate ISDN's two B channels plus one D channel architecture allows at least three simultaneous "conversations" over a standard phone line. Basic Rate ISDN has two B channels for voice, circuit, or packet conversations; and one D channel to carry signals between your equipment and the phone company's central office. In addition to carrying signal data, the D channel can also carry low-speed (up to 16,000 bps) packet data calls.

NOTE What are circuit and packet calls? You'll learn more about this in Chapter 3, but until then, think of them as a train pulling cars (circuit calls) or people on a bus (packet calls). You'll also find more information about the different types of ISDN and communications channels in Chapter 3.

By splitting the line into logical pipes or channels, ISDN can even do two things at once: you can send a fax to one place and receive a fax from another, or utilize a 56-kbps Internet connection while taking a voice call.

Several Devices, Several Numbers, and Multiple Call Appearances

Before ISDN, a phone network had many limitations in the arrangements of service, phone sets, and phone numbers. For example, each line could have only one phone number, and incoming calls could not be directed to individual phones at home. With ISDN, using the proper equipment and configuration, you can connect several physical devices to a single ISDN line. Because ISDN lines are divided into logical channels, up to eight devices—phones, computers, fax machines, credit card readers, cash registers, computers, and more—can be connected on a single Basic Rate ISDN interface, in any combination.

Provisioning for multiple devices reduces the need for additional wiring as new devices are added to a home or business. In addition to simultaneous transmissions, multiple calls can be in progress at one time; in theory, all eight devices could be in use simultaneously. Multiple "call appearances" of one number allow individuals to handle several calls at once or conference them together. An ISDN phone system can support up to 64 call appearances, phone numbers, or IDs of the same or different phone numbers and route incoming calls appropriately.

For example, multiple call appearances are useful for a small office with a large number of outside salespeople. A single physical device, such as an ISDN phone, could accept calls, intermittently or up to two calls simultaneously. Yet each salesperson could have his or her calls answered appropriately by routing incoming calls to individual voice mailboxes. Realistically, a single ISDN phone could handle up to four call appearances. ISDN removes many of analog technology's limitations. When you consider ISDN's multiple-call capabilities, simultaneous transmission, and data capabilities, many new applications are possible. With the possibility of 8 physical devices and 64 call appearances, you gain a great deal of flexibility and a large number of configurations.

ISDN Indicators

We've covered how you can benefit from using ISDN, but you may still be skeptical about whether you'll be able to find the services, equipment, and support you'll need once you've decided to convert your communications system. Here are a few indicators that ISDN has been accepted in the United States and that you'll have no problems making ISDN connections:

- More and more ISDN lines are being installed by telecommunications companies.

- Software companies are incorporating ISDN interfaces into their products.

- The telecommunications industry, including software and hardware manufacturers, is investing more and more in providing ISDN support.

- The cost of ISDN equipment is going down.

The Growing Number of ISDN Lines

The total number of ISDN lines in service is growing rapidly:

- Ameritech claims that it had 20,000 lines at the end of 1995 and will have 40,000 by the end of 1996.

- Pacific Bell has gone from 25,000 lines in 1994 to 60,000 in 1995, with a projection to hit 100,000 by the end of 1996, and perhaps 1 million by 1997.

- Bell Atlantic already has more than 100,000 ISDN lines in service and has recently added residential ISDN service in its territory.

- NYNEX also recently announced increased support for the service and a goal of 1 million lines by the year 2000.

These numbers may be more optimistic than real, but it is very likely that demand for ISDN will create a community of more than 1 million ISDN users by the end of 1997.

The growth in the ISDN service provided by all the RBOCs and carriers will continue to create the momentum necessary to increase ISDN's availability, ease of installation, and multiple-solution packages. You may not even notice ISDN as the communications method when all the line drivers, hardware, and software tools automatically configure your dial-in access to ISDN.

The volume will also spur the internetworking industry to continue to provide ISDN as a standard interface for router, bridges, and file servers. This is another market for interconnecting company and remote sites that is going to continue to expand with the proliferation of client/ server applications. It will not be uncommon for community-of-interest networks or services to use ISDN for smaller, less data-intensive sites' connectivity.

ISDN is not the only network connection option (we'll take a look at the other options a little later in the chapter), but it will be one of the most available and integrated ones. When routers and remote LAN access equipment come with ISDN as a standard interface, then ISDN will become one of the few network options that can be implemented immediately.

Microsoft Joins the Crowd

ISDN support is a standard feature in Windows NT versions 3.5 and later, which means that you do not need any additional software to use ISDN. Microsoft provides the ISDN interface for Windows 95 software. This incorporates some automatic configuration capability and will eventually provide support for a multitude of ISDN terminal adapter choices. The integration of ISDN into the operating system will lead to more corporate LAN access and increased adoption of ISDN as the remote networking connection.

NOTE If you have Windows 95 now and you plan to use an internal ISDN adapter (one that goes inside your PC), you need to get the iSDN Accelerator Pack for Windows 95. This add-on software integrates ISDN with Windows 95. If you plan to use an external ISDN adapter, you do not need the ISDN Accelerator Pack. External ISDN adapters look like a modem to the computer, so no additional software is required. See the chapters in Part 3 for information about specific ISDN hardware and software.

The Microsoft Network will provide ISDN access support for end-user connections. This is happening with all online service providers. As service providers embed Web page browsers and increase the development of graphical user interfaces (GUIs), customers will demand access to higher bandwidth services for additional through-put *and* increased speed for surfing the World Wide Web.

Increasing Industry Support

Never before has ISDN had so much support from the variety of industry players. It has generally been the carriers (and not all of them at the same time) talking about ISDN as some new digital service, without making it clear what that really means in relation to your needs and transition costs.

Today, there are constantly announcements on some expansion of ISDN service, program, or equipment advancement. Some news services have had more than 2000 articles a month on some new development of ISDN service or equipment. The fact that there is so much activity indicates that the industry is finally taking ISDN on as a major communications method. Here are some examples of increasing industry support:

- Intellicom has announced nationwide coverage and multiple carrier support programs for ISDN product packages, offering this support in addition to its support for the Microsoft Network program.

- Tech Data has added more ISDN equipment to its product line and is announcing carrier programs to integrate the entire purchase into a supported end-to-end program.

- *Computer Reseller News* has touted ISDN as the next income generating trend for the reseller and VAR (value-added reseller) industry.

- Shiva Corporation, known for remote LAN access, purchased Spider (for ISDN interfaces) and announced that the company now has an enhanced ISDN product line. This will provide integrated ISDN interfaces in the company's popular LAN Rover products, which many businesses use today.

We have mentioned software and hardware industry players who are providing ISDN interfaces. What is just as important is that they are increasing the functionality of their ISDN product line while they are reducing it in size and cost. These improvements will continue and at a faster pace.

Lower ISDN Equipment Prices

For many years, ISDN terminal adapters were priced at over $1,000, and they came in many separate parts, which made them costly and cumbersome. Now we are seeing equipment options being listed for $375 and even $299. In some promotions by carriers and service providers, the equipment is as low as $199.

TIP

There are many new programs and promotions announced on a daily basis. Watch for new announcements about ISDN promotions and joint marketing programs, such as those offered by Internet providers and carriers. These will provide real cost-effective equipment prices and additional support for ISDN.

By the end of 1996, there will be multiple options available for under $300, and the average cost of ISDN equipment will be under $500. And by the end of 1997, there will multiple equipment options for under $200. This equipment will include many ISDN features, along with an analog interface for backup modem and voice connections. Specific equipment for different ISDN applications is discussed in Part 3.

What about Other Technologies?

There are alternatives to ISDN as an access and service technology, just as there have always been options for the business and home user of communications services. Today, ISDN competes with many network services.

Normal, regular analog dial-tone has been one of the most widespread alternatives and user-friendly options for all types of information transmissions for business and home use. The development of fast modems over the last four years and the availability of the analog dial-tone, voice-designed network has extended the life of modem technology beyond normal expectations. It will always be an option for the ordinary consumer for the next decade.

The following are some of the other new communications technologies, which we will address in the following sections. They are all in the development stages and provide digital connectivity for some of the same applications as ISDN.

- Wireless networks
- Cable modems
- Asymmetrical Digital Subscriber Line (ADSL)
- Wireless cable, both Multichannel Multipoint Distribution Service (MMDS) and Local Multipoint Distribution System (LMDS)

Wireless Networks

Two main types of wireless network technology are becoming more popular with businesses: cellular phone networks and Personal Communications Services (PCS).

Cellular Phone Networks

Cellular technology uses FM radio waves, rather than conventional phone lines, to transmit conversations; hence the reference to "wireless" communications. As you move about with your cellular phone, your call is transferred via a computerized switch between operating areas known as *cells*. Each cell has its own transmission tower linked to a Mobile Telephone Switching Office (MTSO), which connects your call to the public-switched telephone network. MTSOs are owned and operated by one of the two cellular carriers in your area.

When you place a call on a cellular phone, you won't hear a dial tone because the phone isn't linked to your local phone company's phone lines. A cellular phone call is transmitted on a radio frequency to a local cell site. Each cell site is designed to provide coverage to a specific geographic area, generally several miles in diameter. As you move from one location to another, your call is handed off to the next cell site to provide optimal signal coverage and call clarity. This arrangement of multiple cells allows you to travel throughout a territory and maintain a conversation as the call is handed off from cell site to cell site. Figure 2.3 illustrates the path of a cellular communication as the call moves from cell to cell.

FIGURE 2.3:

A cellular communication is transferred from cell to cell.

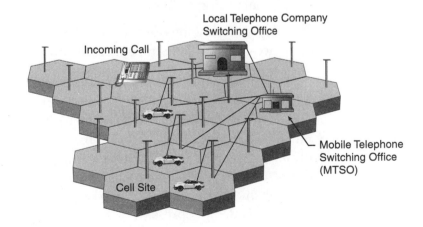

Local Telephone Company Switching Office

Incoming Call

Mobile Telephone Switching Office (MTSO)

Cell Site

The technology most commonly used to transmit your cellular calls is analog. Through analog transmission, your voice is actually carried on the airwaves. The latest evolution in cellular is a new transmission technique using digital technology. Digital transmission, which increases the cellular phone system's capacity, is now being introduced in major metropolitan areas and will eventually be available in most markets.

CDPD, or Cellular Digital Packet Data, service uses some sideband frequency of cellular systems to create a data channel bandwidth that could be sold as a data-packet service. It usually runs at speeds of only 19.2 kbps and requires additional equipment in the cellular network to support the data service. Digital transmission provides several benefits to cellular customers:

- It's more efficient than conventional analog transmission because it handles three calls for every one made in an analog system.

- With digital service, battery life is longer, too; each one provides up to 50 percent more talk time.

- Its increased capacity also means fewer blocked calls for customers and a more private method of transmission than conventional analog technology.

- Digital technology provides a platform for future wireless services, such as data transmission and interactive computers.

The cellular network works well for some interactive data communications, such as short, packet-based exchanges of information (paging, credit card transactions, and so on). This technology may not be reliable enough for long, data-intensive transactions. As the cellular phone system's handling of data is improved and its costs come down, its use will become more common. It will be a good choice for the truly mobile professional worker or service company, such as UPS and Federal Express.

PCS

Most people who know something about PCS have the perception that it is similar to today's cellular telephone service offered commercially throughout the United States. Yet, ask them what is meant by the term "Personal Communications" in the acronym, and you will probably get a blank stare. But personalized communications is the essence underlying PCS.

PCS is a new set of wireless telecommunications services personalized to the individual. This means that phone numbers used in PCS handsets will become tied specifically to an individual, and the types and features of service that each subscriber desires will be customized to his or her unique needs. Thus, the PCS phone number will belong to a person for as long as he or she wants it. Services such as stock quotes on selected companies, voice mail, caller identification, and many others will become specific to the individual holding that phone number. This gives rise to the concept of a personal communications number, or PCN.

The assignment of a unique PCS phone number to a customer will facilitate call initiation and completion across regional, national, and international borders. The network will do all the work of tracking a customer, knowing where he or she is, and facilitating a call through the nearest service operator and cell site. The success of this service, called anywhere call-routing and origination, or seamless roaming, depends on all PCS operators cooperating in providing subscriber information to each other and in establishing agreements to support each other's customers.

NOTE Since there are no other services that compare with PCS, past cellular telephone industry trends have become the basis for projecting PCS future trends. This is done even in some cases where the service provided is significantly different from existing cellular telephone service. Most business proposals include reference to the historical penetration rate of cellular, which is approximately 7 percent after ten years.

In spite of all the references to cellular, PCS should be considered distinctly different. PCS must be viewed in terms of what it offers in vision and in technology that cellular cannot.

PCS has several advantages over existing cellular telephone service:

- Better service quality through the use of digital technology
- More compact radio interface equipment (smaller, lighter handsets)
- Increased mobility
- Enhanced service features (the key concept here is personalized)
- Price (lighter, smaller, less expensive equipment is the bottom line)

More licenses have been purchased recently, but PCS's overall availability is limited and it's more expensive than other options. Its use for data applications is still being developed. Like cellular networks, PCS technology may be attractive to the truly mobile professional.

Cable Modems

Cable televisions systems were originally designed to provide one-way, analog transmission (or *downstream* transmission) of television programming from the cable company's equipment, or *head end*, to the subscriber's home. Coaxial cable has the capacity (*bandwidth*) to support two-way, or full-duplex, transportation of signals, messages, data, and information between the subscriber at home and the cable company's equipment plant. *Full-duplex* is a term used to describe a communications channel that allows data to travel in both directions (send and receive), simultaneously.

TIP

To understand the terms *downstream* and *upstream*, picture the cable company as the source of a stream or river. All information (transmissions) flow like water into the various tributaries (or neighborhoods) and end up in the subscriber's home. Upstream communication is the reverse—information, or requests for programming, Web pages, video, and so on, are sent from the subscriber's home, up the river, collected through the stream's tributaries, and arrive at the cable company's main plant.

An explosion of services and information on the information highway and World Wide Web has created a demand for Internet access and other two-way services, pushing the cable industry to enhance existing cable systems with fiber-optic technology. By upgrading the cable plant, or head end, cable companies are transforming their networks into hybrid fiber-coaxial (HFC) networks. By replacing this part of the distribution plant with fiber-optics, the cable companies are converting their networks and adding the capability for two-way transmission. Several companies expect to complete the transformation of all of their respective systems as early as 1998, and the transition may be completed even earlier in metropolitan areas.

How Cable Modems Work

Similar to ISDN services, cable modems have been talked about for the last five years, but have only recently been embraced by a number of well-known equipment manufacturers. This technology is still being tested for reliability and full-duplex transmission. Most methods involve different upstream and downstream communication streams. For example, some are capable of 6 to 10 megabits downstream and 768 kilobits upstream.

The cable modems convert digital information into a modulated radio frequency (RF) signal and convert RF signals back to digital information when they arrive at their destination. The conversion is performed by a modem at the subscriber's home, and again by the head-end equipment that handles multiple subscribers at the cables main plant. Figure 2.4 shows the skeleton or the tree topology of a cable company's coaxial distribution from the company's head end to individual subscribers in a neighborhood.

FIGURE 2.4:

The "bones" of a cable company

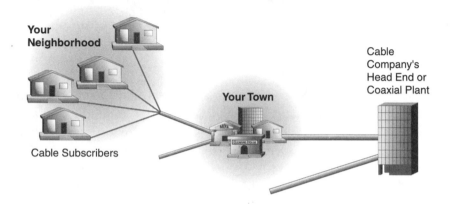

Cable Modem Technology Hurdles

Cable modems can provide the service for Internet access and cable-to-cable subscriber service on the same cable system. However they cannot provide the communication for any-to-any connectivity of a switched service, which is offered by both analog dial-tone and ISDN.

The cable industry must gear up to support this new digital access service. First, the companies must find the capital and implement the system, and then they will need to invest in the expansion of their services while they also try to enter other new businesses. It is somewhat similar to the ISDN upgrade to the switched network.

Currently, the cable network systems cover less area than the phone networks. It will be some time before cable modems are nationally available, with a high service level and standard offerings. Experts have said that the industry needs to clear several obstacles beyond mere modem availability before high-speed data services can become a mainstream business. Among them are the reverse-path activation (upstream) needed for bidirectional signal flow (full-duplex), customer service installation, cable-modem standards, contention, security, costs, and billing strategies.

In many cases, users need to upgrade their systems to provide cable modem service. In all cases, the cable modem coaxial connection or jack must be wired separately and in front of connections to television connections. If the cable signal is weak, it will need to be boosted.

Also, the tree topology of the cable arrangement creates three technical hurdles:

- **Contention:** With contention, everyone in the neighborhood shares the same connection, or bus. For example, with cable modems, a speed of 10 Mbps is common. However, if you assume a 10 percent take rate in a neighborhood of 500 homes, then each user gets only 200 kbps—not much more than ISDN.

- **Security:** Because users or cable subscribers share the same connection, it's possible for everyone in the neighborhood to listen in on every transaction.

- **Ingress:** Another problem is ingress (or noise) on the cable company's head end, or coax plant. The frequency band that the data is transmitted over, 5 to 40 MHz, is ideal for picking up AM radio stations, ham radio broadcasts, and even household noise. In addition, the tree topology focuses noise from every location to the head end, where all the signals (and noise) meet.

There are now multiple cable systems and cable providers with multiple implementations. Today, there are least three different major implementations and vendors: LANCity, Motorola, and Zenith.

The CyberSURFR, Motorola's first-generation of cable modem, offers throughput speeds of up to 10 Mbps in the downstream path and 786 kbps upstream. Zenith's current cable modem, the Homeworks Universal, provides 4 Mbps both downstream and upstream.

There are several other cable modem manufacturers, including Toshiba and Hewlett-Packard.

Currently, cable companies are setting expectations at no more than $30 to $50 per month. Will consumers double their cable bill for this? Also, the customer premise equipment (CPE—for example, a telephone is a CPE) is expensive and requires a NIC (network interface card) because the cable network is Ethernet, a shared connection.

Although these technical problems are not insurmountable, they are not trivial. Maybe we should rename this section, "Things Your Cable Company Never Told You and the Questions They Are Afraid You Will Ask."

A Closer Look at Cable Modem Developments

If you think that telephone-independent companies and RBOCs working together for providing a standard ISDN service is a feat, try to imagine multiple cable systems across the nation upgrading and standardizing on cable modem service in a coordinated fashion. However, during the last few months, we've seen many announcements and developments regarding the cable modem technology. It may have finally broken out of the "hype" marketing phase.

Many of the cable giants have announced multiple field trials of cable modems. Time Warner, Continental Cable, and TCI have made recent specific trial programs public. Continental Cable has also announced a trial in the Boston area with 200 users, using LANCity equipment. Most of the trials include some prototype equipment and require infrastructure upgrades to the cable system. The trials are intended to identify cost and technical service issues.

Right now, there is a lot of press about $10 a month for cable modem service. A service close to Ethernet speed for this price is an attractive option.

The plan is to eventually make cable modems part of the digital upgrade of cable systems over the next few years. The cable companies will need to weigh the costs against all the over investment options. The move to digital, the HFC (hybrid fiber-coaxial) upgrades, and the telephony capital investments will all diffuse the cable industry's move to provide a national cable modem infrastructure. There will be competition for the investment capital for these separate services, which will slow the cable modem developments.

The cable modem technology will need to be integrated into PC or terminal equipment, going through the same maturation and cost improvement process that ISDN terminal adapters have. This will not happen until cable companies have created a standard, implemented it, and made interconnection agreements.

Some manufacturers and equipment vendors, like Intel, Microsoft, and Motorola, are hedging their equipment bets with this technology. Motorola recently announced a package that will enable cable companies to upgrade their cable systems plant to allow for cable modem access and digital connectivity for their subscribers. Time Warner and TCI have made commitments to the Motorola cable modem equipment (for over 250,000 cable modem units).

Although these companies come at this business from the PC or terminal equipment side, they all want to be part of the internetworked PC or workstation in a business or home. There is still the dream of the "magic" box in the home that will provide computing, communications, and entertainment needs of the United States households. Investing in this development is valuable to the major players in this market.

continued on next page

The cable modem option will be viable and may have a different customer set than ISDN. With cable systems being more residential in nature and phone systems being directed toward business customers, there will be room for both services. For the next three years, ISDN will have a window of opportunity for increased market share. After three years, it will come down to availability and affordability. Even with all the unknowns, it is clear that cable modems will be an option, and there will be room for both ISDN and cable modems until the year 2000.

ADSL

ADSL is another alternate communications technology that has been getting some recent attention, although it also has been around for more than five years. It has been sponsored by some RBOCs and carriers, but only three RBOCs (Bell Atlantic, NYNEX, and Southwestern Bell) are really serious about this technology at this time. As with other new types of communications technologies, the companies involved in its development are finding it difficult to juggle the copper cable plant requirements while increasing fiber-optic and HFC networks to the home.

ADSL can be used to deliver high-rate digital data over existing ordinary phone lines. A new modulation technology called Discrete Multitone (DMT) allows the transmission of high-speed data. ADSL facilitates the simultaneous use of normal phone services, ISDN, and high-speed data transmission, such as video. DMT-based ADSL can be seen as the transition from existing copper lines to the future fiber-optic cables. This makes ADSL economically interesting for the local phone companies. They can offer customers high-speed data services even before switching to fiber-optics.

About ADSL

ADSL started out as the phone company's way to compete with cable television by delivering both television and phone service on your plain old copper phone line. Now it's also a good candidate for high-speed Internet access.

The *A* stands for Asymmetric, meaning the phone companies can send lots of data to you, but you can't send much to them. This is close to perfect for Internet connections, since you receive lots of graphics and data from today's World Wide Web sites, but send back a tiny URL (for Uniform Resource Locator, that's what the http:\ stuff is called). Originally, only 16- or 64-kbps uplink (what you send) was supported; recent flavors (or varieties) of ADSL support up to ten times that much!

ADSL is one member of a continuum of last-mile transport systems called DSL, or Digital Subscriber Line, which can carry about 1 to 6 Mbps over copper lines. It does not include any way to make long-distance data calls, or even local calls; in other words, no phone service. However, that is another matter entirely—and still being debated. ADSL was originally designed for use as a regular phone line when the power goes out, which would be an improvement over Basic Rate Interface ISDN.

Currently, the only form of DSL really being deployed is HDSL (High-bit-rate Digital Subscriber Line), which may be considered a direct replacement for traditional T1 service (more about T1 in Chapter 3). T1 lines have been around for years, but require technicians to tune the line to perfection; HDSL modems can handle noisy copper lines easily, so HDSL may be less expensive to install and run. In general, the fastest DSL schemes go only a couple miles, but slower DSL modems will go farther—just one more technology trade-off.

This type of ADSL requires two pairs of copper wires to the facility. Since most homes are wired for one or two phone lines, residential customers would need to install another set of wires to use ADSL.

Also, as mentioned, ADSL provides no phone service, so you'll want to keep your current phone service. Another option is for two-wire ADSL technology, which provides less bandwidth but does make the service more available.

Testing, trial runs, and product developments will dictate how viable an option ADSL will be compared with ISDN service. It will take some time to turn the resources around and either split or divert ISDN resources to ADSL service in the industry.

Are DSL Modem Costs Coming Down?

PairGain claims that the company will be conducting field trials of consumer DSL gear this year! And it already has a distributor for its Campus products. PairGain promises a home version will be available for $995 soon, but the Campus version is available now; a distributor is quoting a minimum of $2,500 for a stand-alone unit.

PairGain claims a maximum range of 8 miles, but that's only at the lowest speed, 384 kbps, and with the best wire. The distance coverage depends on the gauge of the installed copper and the transmission speed. At higher speeds or with thinner wire, the maximum range can drop below 2 miles. The "hard-core" or the rich can use these to surf the Internet now, if you meet the following criteria:

- Your ISP must offer a high-speed connect option.

- You can order a special setup or private network from your ISP through your phone company's central switch then to you.

- You have $600 to $1,200 per month (around or in the ballpark of the cost of a T1 connection) to spend on your ISP's high-speed connection option.

- The total wire length from your ISP through your phone company's central office *then* to you is about 2 miles (5 miles for 384 kbps).

- You have at least $6,000 to spend on equipment ($2,500 for each DSL modem, and at least $1,000 for routers or high-speed serial ports).

As you can see from the above requirements, ADSL may be used to supplement other technologies, but it is not likely it will be pursued as the mass-market solution for home access and connectivity. One exception may be in remote areas where ISDN is more expensive to provide and the ISDN switch resources are not available. It is more likely that this will be an interim technology until HFC networks are made available and more economical.

MMDS Wireless Cable

MMDS technology currently allows operators (wireless cable companies) to send up to 33 analog channels of wireless cable to subscribers.

MMDS stands for Multipoint Multichannel Distribution System, and is also called wireless cable, as in cable television without the cable. Coaxial cabling is the most expensive way to provide an area with television programming. Cable companies install tens or hundreds of thousands feet of expensive coax that risks being damaged or stolen, just to get television programs to the subscriber. Wireless cable, or MMDS, is a new solution. With this technology, a complete package of television programs can be transmitted on a very high frequency (2.5 to 2.7 gigahertz). Because microwave signals travel only in a straight, line-of-sight fashion, the transmission antenna height is very important; it must be as high as possible. Generally, MMDS systems cover areas between 12 and 30 miles. MMDS signals are normally encrypted, to allow the operator control over billing and services.

At the reception side, MMDS uses a small receiver. The receiver can be a small antenna or a small parabolic reflector, depending on the distance between the transmitter and the subscriber's antenna. At the antenna, a converter changes the microwave frequencies into the normal UHF of VHF signals found on all televisions. Because MMDS transmitters are being converted to modulate in the same way as normal VHF/UHF transmissions, no extra demodulation is necessary. In an encrypted MMDS system, a decoder is placed between the antenna and the television set. Depending on the encryption scheme, the decoder can address each channel, or a complete package can be decoded simultaneously.

MMDS technology offers a quick way to provide cable services in a metropolitan area, particularly to those without an existing cable network. It does not yet support digital access and two-way communication for data transport.

MMDS has been around for years, but has been relegated to niche markets. It was primarily considered a cheap way to get analog channel capacity to rural areas. With an average of 25 analog channels available in each licensed area, it does not provide much competition in the cities and suburbs where cable companies offer more than 60 analog channels. But with the advent of compression technologies, these 25 channels became 125 channels, allowing MMDS to compete with traditional wireline cable and direct satellite systems, especially in urban areas where there are more potential subscribers per transmitter.

Meanwhile, the success of DirecTV has begun to drive down the digital setup prices to make wireless digital television affordable. DirecTV has sold one million receivers that include an MPEG decoder, and Digital MMDS can build on this substantial installed base. DirecTV has also played a role in convincing cable and phone companies that there's some quick money to be made in digital near-video on demand (NVOD) systems without having all those expensive wires and switches to install. Although fiber-optics and coaxial still dominate their long-range plans (due to their greater capacity), and while some cable companies are aggressively pushing the roll-out of digital HFC networks for NVOD, for the phone companies (who don't already have an investment in coaxial to the home), the next few years belong to wireless.

MMDS has been drawing attention in the telecommunications industry. In recent months, the three RBOCs investing in the TeleTV programming consortium (Pacific Bell, Bell Atlantic, and NYNEX) have also invested in digital MMDS (wireless cable) companies. They see the technology as a means of providing large numbers of video channels quickly in their franchise areas.

Recently, companies like Bell Atlantic, Philips, and Texas Instruments have been showing interest in a different wireless "fiber-in-the-sky" technology: LMDS.

LMDS Wireless Cable

Local Multipoint Distribution System (LMDS) uses super-high frequency microwaves to send and receive two-way broadband signals in an area, or cell, approximately 3 to 6 miles in diameter. Covering less than half the area of MMDS, LMDS looks a lot like the narrowband operations of cellular telephone systems. However, the video, voice, and data broadband LMDS signals are capable of two-way transmission.

NOTE
Narrowband refers to sub-voice channels able to carry data starting at speeds up to 64 kbps, ranging up to T1 rates. This term is sometimes used to refer to POTS and non-video-capable systems. *Broadband* refers to a transmission medium capable of supporting a wide range of frequencies, typically from audio up to video frequencies. It can carry multiple signals by dividing the total capacity of the medium into multiple, independent bandwidth channels, where each channel operates only on a specific range of frequencies. Broadband systems can typically deliver multiple video channels and other services. Coaxial cable TV networks are a classic example of broadband services, with numerous video channels. We'll discuss broadband in more detail in the next chapter.

Microwaves have long been used to transmit data, but the conventional wisdom believed there were limits; they could be used for point-to-point communication and little else. Consequently, a section of the RF spectrum between 27.5 and 29.5 GHz (gigahertz, or billions of cycles per second) went virtually unused except in outer space.

Until recently, everyone agreed that using high-frequency microwaves to transmit video was impractical. Engineers had been focusing their attention lower in the RF spectrum, because lower-frequency signals (with enough power) can be sent long distances and can move through buildings.

Additionally, most people with a microwave oven understood that microwaves are absorbed by water, exciting the water molecules to heat up and boil. Rain would literally wash a video signal off a screen, by reducing the strength of the desired signal and adding unwanted noise to the microwaves.

By using Frequency Modulation (FM) rather than the standard Amplitude Modulation (AM) used in cable systems, LMDS signals could be generated with ten times higher signal quality (20 decibels) than cable; thus creating a surplus of signal that allows the signals to pierce even hard, driving rain. Also by combining a very high frequency and small cell sizes (3 to 6 miles in diameter), LMDS requires only a small, 6-inch square antenna to receive the signals. Currently, the patents for LMDS technology are held by CellularVision Technologies and Telecommunications (CT&T), which plans to license the technology.

This technology has the limitations that are native to any line-of-sight technology like microwave transmission. It also is less available than the other communication system options.

Like the other new communications being developed, MMDS and LMDS are now in the testing and trial run phases. It may take some major development and industry investment to come up with a viable service and business model for this technology, even without the data service. Then there is the work to provide the end-to-end solution of equipment and support for full-duplex service that has yet to be developed, tested, and marketed.

When You Should Consider ISDN

ISDN should be part of your thinking as soon as you, as an end user or a corporate MIS manager, see the following conditions:

- End users demanding access to corporate resources on the LAN
- Frustration with the analog/modem access speed
- A growing number of workers (part-time or full-time) working from their homes
- Executives asking for the access to multiple corporate sites
- Frequent access to online services and/or e-mail
- More and more large data file transfers
- Branch offices who need on demand access to corporate headquarters
- Users looking for data collaboration or video connections

These are just a few of the conditions or trends that indicate that you need to take a look at ISDN as an alternative for network connectivity. You'll find more details about specific applications later in the book.

CHAPTER
THREE

3

An ISDN
Anatomy Lesson

- Telecommunications terminology

- Types of ISDN service

- ISDN channels: B, D, and aggregates

- ISDN connections for internetworks

- ISDN physical connections

- Phone company's bearer services for ISDN

- National ISDN and other standards

Now that you know what ISDN is and why it's a good choice for data and voice communications systems, let's take a look at how it works and how it is put together. The information in this chapter will help you understand what you need for your own ISDN system. First, we'll review some general telecommunications concepts and terminology that are associated with ISDN services. Then we'll take a look under the hood and see how all of the logical and physical pieces of an ISDN interface fit together.

A Telecommunications Primer

To understand how ISDN works, you need to know about some of the methods and terms used in digital telecommunications. The following sections describe some basic elements of telecommunication systems:

- Communications protocols—TCP/IP, SLIP, and PPP
- Data-transmission methods—circuit-switching, packet-switching, and the X.25 protocol
- Digital Services—DS0 through DS4

Communications Protocols

A communications protocol is a set of guidelines that regulate how two or more end points communicate with each other.

Communications protocols allow data to be transmitted over communication channels.

In telecommunications discussions, you'll hear about the PPP and SLIP protocols. PPP and SLIP are both methods for transmitting TCP/IP frames over a serial connection (such as a modem). Most people are using PPP, which is widely supported by the current software.

What's TCP/IP?

TCP/IP is one more pesky and cryptic acronym for the protocol suite used to move data over the Internet. IP stands for Internet Protocol (often referred to as the Internet address), and TCP stands for Transmission Control Protocol. TCP is really built on top of IP, but nearly all Internet services use TCP to transfer data because it is more reliable and easier to use.

TCP/IP transmits data by breaking it down into *frames*, or limited-size units of data. Each frame has a header with information about the frame size, originator, destination, and other identifying information, followed by a chunk of data. This is different from your standard text account, which sends the characters you actually see and type over your modem connection.

What's SLIP?

SLIP is an acronym for Serial Line Internet Protocol. Basically, it is a protocol that communicates between the IP address of a host server (such as a World Wide Web site) and a client (that's you) over a serial line. During communication between the client and server, SLIP takes packets through the serial line and translates them for the client terminal.

SLIP was originally intended as a cheap way to network computers together. All SLIP does is move IP frames over a serial connection. It does not contain any built-in methods for setting up and configuring a connection, since it was designed for permanent or dedicated connections.

SLIP became increasingly popular as modems became increasingly faster and the data could be transferred more rapidly over phone lines. It provided a simple way to access a server from a remote computer and use resources without being physically (or directly) wired to the server.

A variation of this protocol is CSLIP, which stands for Compression Serial Line Protocol. CSLIP communication is also emulated over a serial line, but the packets transmitted are compressed by the server before being sent to the client terminal. When the packets reach the client, the packet is uncompressed and passed into the terminal's storage. This process accomplishes two things: it reduces traffic over the network and the serial line, and it reduces the download time from server to client.

What's PPP?

PPP stands for Point-to-Point Protocol. This is another protocol that gives the client terminal access to the resources of a server without being directly physically connected. Like the SLIP connection, PPP can run over phone lines and encode the information before sending it to the recipient.

Although both PPP and SLIP connections enable a distant user to connect and access resources of a remote server over a serial line, they handle the data differently. Because PPP has many distinctive advantages, it is more widely used than SLIP.

PPP is a more recent protocol that is designed to work well for dial-up modem connections, and it sets up your connection faster than SLIP. PPP has subprotocols that obtain the information needed to set up a TCP/IP connection automatically; SLIP does not. PPP also incorporates the header compression method found in CSLIP.

WARNING The terms PPP and SLIP are sometimes used interchangeably, but they are not the same. Devices using SLIP and PPP cannot communicate with each other or share data. If you run SLIP on your host PC, the server must support connections running SLIP. The same applies to PPP.

One big difference between the two protocols is the networking protocols necessary to emulate the connection. In SLIP connectivity, the network must run TCP/IP as its primary communication source between resources. PPP connectivity will run on a wide variety of protocols: TCP/IP, NetBIOS, and IPX/SPX.

Another difference between SLIP and PPP connectivity is the number of concurrent communication applications supported. A SLIP connection allows only one communication application to be active at a time. PPP supports multiple applications.

Although there are differences between PPP, CSLIP, and SLIP in terms of how they operate, your protocol choice will not impact the kinds of network software that you can use. Any network program (such as a mail reader, news reader, or file-transfer program) will talk to the TCP/IP driver software that you install (such as Trumpet Winsock MacTCP or the software included in the Macintosh and Windows operating system). The TCP/IP driver will furnish a standard interface that other programs will use to establish network connections and send data.

Switches and Switching

When you make an ISDN or a POTS call, you connect to a phone company switch. This switch is basically a powerful computer specifically designed to route calls. There are many different switch types in use with ISDN. The two most popular switch types are the NT and AT&T 5ESS switch. These switches are very similar in functionality, but the AT&T has a few more features. Unfortunately some of these switches are incompatible with each other. To resolve this problem, a national standard, called National ISDN, was created. This standard allows different switch types to communicate with each other. ISDN standards are covered later in the chapter.

Circuit-Switching: Routing Voice or Data

Circuit-switching is the process of setting up and keeping open a circuit between two or more users. The users have complete and exclusive use of the circuit until it is closed. A telephone call is an example of circuit-switching.

In a circuit-switching technology, a direct connection is created through the switches that reside between the communicating stations. Originally, it was designed for voice traffic, which needs a dedicated line for conversation between two people. However, the two users do not have direct wires through a circuit-switched network. Instead, the intervening switches have electronic connectors that "couple" the communications links directly to each other.

Circuit-switched voice (CSV) service is a digital voice service that offers many of the capabilities of a business Centrex, such as call-waiting, speed-calling, and call-transfer, over an ISDN Digital Subscriber Line (DSL). Traditionally, phone services are CSV. In this service, a voice must go through several switches before reaching its final destination.

For point-to-point data connections, you need to use circuit-switched data (CSD) service. This type of service is designed to carry digital data through the switches. CSD service provides end-to-end digital service to pass data or video information over the public network.

Packet-Switching: Routing Multiple Data Packets

Packet-switching is another data transmission method. Phone services can also be packet-switched data (PSD).

Packet-switching is so named because a user's data (such as a message) is separated and transmitted in small units, called *packets*. Each packet occupies a transmission line only for the duration of the transmission; the line is then made available for another user's packet. Packet size is limited so that packets do not occupy the line for extended periods.

With packet-switching, a channel is occupied only for the duration of transmission of the packet. Packets are independently transmitted from point to point between source and destination and reassembled into proper sequence at the destination. Packet-switching has become the prevalent switching technique for data-communications networks. It is used in such diverse systems as PBXs, LANs, and even multiplexers (devices that allow the transmission of a number of messages simultaneously over the same communications channel). A packet-switched network uses multiple routes (paths) between the switches within the network. The packets are routed across the paths in accordance with traffic congestion, error conditions, the shortest end-to-end path, and other criteria.

NOTE The packet-switching network topology is different from message-switching in networks. A message-switching system passes messages from sender to receiver through intermediate nodes. Packet-switched networks use more switches, allowing the traffic load to be distributed to other switches. At least three, and often more, trunks are attached to the switches. This arrangement allows the network to route the packets around failed or busy switches and lines. Consequently, a packet-switched network gives the user better availability and reliability than a message-switched network.

In circuit-switched (telephone) systems, the time to set up a connection is often lengthy (sometimes, several seconds). In contrast, packet-switching systems use dedicated leased lines or dial-up connections, which are immediately made available to users.

When packet-switching is used with ISDN, it is usually for the D-channel data. Using your D channel, it is possible to implement various low-bandwidth services for communicating with other ISDN users. Packet-switching can also be used on the B channels, although this is generally for X.25 or similar networks.

The X.25 Protocol for Packet Connections

X.25 is a CCITT (now ITU) standard that defines the packet format for data transfers in a public data network. These networks can be used for other types of data, including IP, DECnet, and XNS.

The X.25 protocol defines the procedures for establishing and maintaining a data link between two network components. This protocol defines the network procedures used to initiate, maintain, and release packet connections. Packet services can be divided into two types of connections:

- *Connectionless service*, also known as *datagram service*, places data packets onto the network independently. Each packet has the address and service information required to route the packet through the network to the destination node. In this type of connection, packets may take a number of different routes to reach the correct address. Different routes may be required due to network congestion and/or link or component failures.

- *Connection-oriented service* retains connection information until the completion of the connection session. All packets involved in the session are routed via the same nodes to the address location. This is the type of packet service provided with ISDN in the 5ESS switch.

NOTE
> The X.25 protocol defines procedures that were developed in the 1970s. These procedures provide for extensive error detection and data retransmission, because the state of the public network at that time required such robust procedures. The present-day public network is primarily a digital network and much less susceptible to data loss due to noise and link failure. Many applications that have used X.25 in the past may now be more readily served via frame relay or ATM (Asynchronous Transfer Mode).

The X.25 protocol also defines two types of packet connections:

- A *permanent virtual circuit* is a connection that is permanently assigned. Whenever you send information over a permanent virtual circuit on the network, it is always delivered to the same terminal, because the recipient's address is always the same. The circuit is "virtual" because it appears to be a hardwire connection between the two terminals, although one does not exist.

- A *switched virtual circuit* is a circuit where the information is delivered based on the address information provided on a session-by-session basis. In other words, address information must be provided each time a new session is initiated.

Digital Signal (DS) Services

In North America, DS is a telecommunications service that defines a four-level digital-transmission hierarchy, with increasing bandwidths. The hierarchy defines protocols, framing formats, and other specifications for each level.

Bandwidth refers to the amount of data that can be sent through a given communications circuit per second, or its transmission capacity. Specifically, it means the range of frequencies a media (such as copper wire, coaxial cable, satellite, or fiber-optic cable) can effectively transmit. Generally, the larger the bandwidth, the greater the capacity of voice, video, or data the media can carry. Bandwidth in systems is limited both by the transmission media and by the electronics sending and receiving data.

The data signals are transmitted over T-carrier lines, such as T1 or T3. In digital communications, T1 is the carrier used in North America. T1 has a bandwidth of 1.544 Mbps, made up of twenty-four 64-kbps channels. The individual 64-kbps channels are known as DS0 channels. The higher-capacity channels are based on the 64-kbps DS0 channel. A T3 channel has a bandwidth of 44.736 Mbps, and is the equivalent of 28 T1 channels, each of 64 kbps. Table 3.1 shows the DS levels. In Europe, a similar system of E-carrier lines is used. E1 has a bandwidth of 2.048 Mbps. E3 has the highest transmission rate—34 Mbps—generally available in the European digital infrastructure.

TABLE 3.1: Digital Signal (DS) Hierarchy of Channel Capacities

Level	Description
DS0	A single 64-kbps channel of a DS1 digital facility
DS1	A 1.544-Mbps (U.S.) or 2.048-Mbps (Europe) digital signal carried on a T1 facility
DS1/DTI	Domestic trunk interface circuit to be used for DS1 applications with 24 trunks
DS2	A 6.312-Mbps digital signal
DS3	A 44.736-Mbps digital signal carried on a T3 facility
DS4	A 274.176-Mbps digital signal

ISDN Interfaces: Which One Is Right for You?

In ISDN, the conventional analog phone signal is replaced by a 64-kbps digital data stream, called a B channel. In an ISDN voice call, this B channel carries the digitized speech; in internetworking applications, we use it to carry data. To make, receive, and generally control calls, an additional signaling channel called the D channel is used.

ISDN is available with two main types of interface: Basic Rate Interface (BRI) or Primary Rate Interface (PRI). The difference is in the number of B channels that each can support. Broadband ISDN (B-ISDN) is another, faster type of ISDN service being developed. These types of ISDN are described in the following sections.

BRI for Multitasking

The beauty of ISDN is that it can be built in various sizes, not one size fits all like the analog system. The single size, or "compact, " version is BRI.

BRI is the service that corresponds directly to the conventional phone line installed in a home or small office. It is delivered over a pair of copper wires, and the ISDN plug is similar to a normal phone jack.

BRI is the most appropriate type of ISDN service for computer connections and individual use. BRI divides the phone line into three digital channels: two B channels and one D channel, often referred to as 2B+D. As we've said, each of these channels can be used simultaneously, so you can perform several communications tasks at the same time.

The B channels are used to transmit data at rates of 64 kbps or 56 kbps (depending on your phone company). With two B channels, you can make two calls simultaneously. The D channel is a slower, 16-kbps channel for signaling. This channel does the administrative work, such as setting up and tearing down the call and communicating with the phone network. This service offers a total bandwidth of 128 kbps, meeting the needs of most individual users. We'll go into more details about the ISDN channels a little later in the chapter.

WARNING In the United States, you might discover that your local phone company's ISDN service includes only a single B channel, plus a D channel, with a second B channel for an additional charge. The extra charges may feel like information superhighway robbery, but unfortunately, the real reason is that your phone company needs to update its old equipment.

BRI is designed to carry the most data possible to the home through existing copper phone lines (twisted-pair cable). Some time ago, the phone companies discovered that you could squeeze about 160 kbps into those "tiny" copper lines. As shown in Figure 3.1, using ISDN BRI, the phone company can provide two B channels (64 kbps + 64 kbps), and one D channel (16 kbps), and still have 16 kbps for the overhead (data framing, maintenance, and control) of communicating with your house's phone network.

FIGURE 3.1:

The phone company divides BRI transmissions into three logical pipes (two 64 kbps B channels and one 16 kbps D channel) with 16 kbps left over for overhead.

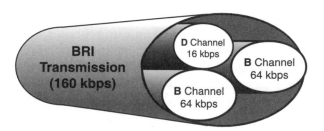

You may find that your phone company offers B channels that carry only 56 kbps. Currently, older phone equipment assumes that analog data is being transmitted. In an analog circuit, extra bits are pulled out from the higher frequencies of the audio in order to do in-band signaling. (Auditory perception is between 20 Hz and 20,000 Hz, but speech normally falls between 3000 Hz and 5000 Hz, so stealing from higher frequencies, was not a problem.) In ISDN circuits, the signaling belongs on the D channel.

NOTE   *Out-of-band signaling* uses frequencies outside the normal range of information-transfer frequencies. This is in contrast to *in-band signaling*, which is transmission within a frequency range normally used for information transmission. Another term for in-band signaling is *bit-robbed*.

Until the phone company upgrades its equipment, the signaling takes some bits out of each B channel, so you get 56 kbps instead of 64 kbps. The 56-kbps limitation is only between switches, and the problem is not limited to the RBOCs. Most interexchange carriers (IXCs), or long-distance companies, cannot provide clear 64-kbps channel service everywhere until they upgrade their T1 systems.

PRI for Big Data Transfers

PRI is designed for businesses that need to transmit larger amounts of data or want to set up their own local phone system. PRI is generally just a faster connection to the phone company. It's useful for intensive bandwidth-on-demand applications, such as LAN interconnections and videoconferencing. (In *bandwidth-on-demand* applications, when data is being sent, the connection is open; when the line is idle, the connection is automatically closed.)

PRI consists of an aggregation of multiple B channels. In North America and Japan, the most common PRI service uses 23 B channels and 1 D channel (64 kbps), which is the equivalent of the DS1 service in the United States (see Table 3.1). Notice that the D channel in the PRI version is "heftier" at 64 kbps than the D channel in the BRI version at 16 kbps. In other countries, the PRI has 31 user channels, usually divided into 30 B channels and 1 D channel, and is based on the E1 interface. Figure 3.2 illustrates PRI components.

FIGURE 3.2:

Primary Rate Interface (PRI) service in the United States

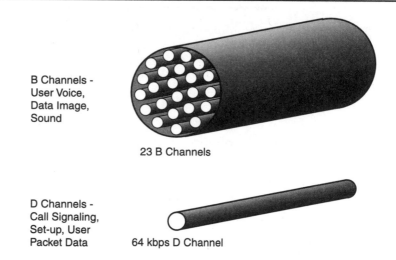

B Channels - User Voice, Data Image, Sound

23 B Channels

D Channels - Call Signaling, Set-up, User Packet Data

64 kbps D Channel

PRI is intended for larger commercial users, and this service is generally attached directly to a PBX or other high-volume, data-networking product. Since there is a substantial difference between the bandwidth of BRI and PRI, many telecommunications service providers offer intermediate services which are a subset of PRI. With PRI, you also have the option of combining several B channels into one "big fat" channel or logical pipe, known as an H channel. These intermediate services are described later in the chapter, in the section about aggregating channels for higher bandwidth.

Broadband-ISDN (B-ISDN): The Next Step

Broadband ISDN, or B-ISDN, is another, faster type of ISDN service currently being developed. B-ISDN is a digital service with speeds in excess of 1.544 Mbps.

B-ISDN is the service of the future. The higher speeds offered are required to support the many applications of the "information super-highway." The speeds for these services usually range from 25 Mbps up to the gigabit range. The two speeds that are most often discussed are OC 1, which is 155 Mbps, and OC 3, which is 622 Mbps.

The speeds in B-ISDN are made possible by the high quality of the digital facilities in place on the network. The early data protocols, such as X.25, required extensive overhead to ensure the delivery of data. Error correction and flow control were performed at a number of intermittent points along the way of a data connection. The new digital facilities and the introduction of fiber-optics have all but eliminated this need.

The B-ISDN service can be in several forms:

- **Frame relay service:** Frame relay is a connectionless service, meaning that each data packet passing through the network contains address information. Even though the most used speeds for the service are currently 56 kbps and 1.55 Mbps, frame relay is considered to be a B-ISDN service. One of the unique facets of frame relay is that the service supports variable-size data packets.

- **Switched Multimegabit Digital Service (SMDS):** SMDS is a digital service that provides a high-speed digital path. The transport speed for SMDS is usually 155 Mbps.

- **Asynchronous Transfer Mode (ATM):** Also known as *cell relay*, ATM is a flexible service made possible by the size of the packets (cells). Typically, the transport speed of most ATM applications is 155 Mbps, with the promise of 622 Mbps in the near future. The cell size for all applications is 53 bytes. The small cell size allows a variety of applications to run on ATM networks, including voice, video, and data.

Many consider ATM to be the transport service of the future. The appeal of the service has to do with its ability to pass voice and video information. These two services are time-sensitive, otherwise known as *isochronous* data. This means that voice and video are susceptible to time delays. The small cell size of ATM and the service options, such as Continuous Bit Rate Service, allow this traffic to flow over the network. Frame relay and SMDS cannot guarantee this level of service.

The Transmission Pipe: The ISDN Channels

Now let's take a closer look at the two ISDN logical channels: B, the bearer channel and D, the signal channel.

The channel names are allegedly from analog circuits being called A channels (A representing analog). The next type of channel was logically labeled B, this is often referred to as the bearer channel. The D channel was labeled the delta channel, because of the channel's relationship with the B channels. However, typing a Greek symbol was troublesome, so it became simply D.

B or Bearer Channel

The basic building block of ISDN is the B (bearer) channel. The B channel is known as the bearer channel because of its role in the transmission system. The B channel is designed for voice, video, or data. It can support synchronous, asynchronous, or isochronous services (or transmitting at the same rate, not at the same rate, or at the same time, respectively) at rates up to 64 kbps. B channels can be aggregated (combined) for higher-bandwidth (speed) applications.

The transmission rate of 64 kbps springs from the fact that this is essentially the data rate at which the analog lines are sampled at (8000 samples per second, 8 bits per sample) for the phone company's Integrated Digital Network (IDN).

Since a B channel supports up to 64-kbps throughput, it lends itself well to higher-speed PC-to-PC file transfers. Using multiple channels and compression, it is possible to approach file-transfer speeds that are found on LANs. The B channel also supports collaborative application sharing, screen sharing, central site synchronous services, and access to remote LANs.

D or Delta Channel

The D (delta) is a 16- or 64-kbps channel that handles the communications between the phone switch and the ISDN device. The BRI D channel provides a 9.6-kbps partition for user data. The D channel supports packet data (the type commonly used in PC serial ports) packet data.

In addition to supporting signaling, the D channel provides a packet service that does quite well with PC-to-PC file transfers. The D channel also supports X.25, a protocol that handles multiple logical sessions on a single channel. It can provide economical host access by allowing many users to share a common ISDN line.

You can think of the D channel as ISDN's traffic cop, signaling, messaging, and keeping the voice and data "traffic" moving smoothly. There are two different types of signaling used in ISDN:

- Digital Subscriber Signaling System 1 (DSS1)
- Signaling System 7 (SS7)

DSS1

ISDN uses DSS1 to communicate with your local phone company. DSS1 acts as a guide for data, defining the format for outgoing data into the pipe(s), addressing, and finally formatting the incoming data. In addition, the D channel handles messaging; for example, establishing, maintaining, and dropping transmissions.

Once your DSS1 signal reaches your local phone company, the company's own signaling system takes over to pass call information within its local system and among other carriers.

SS7

Local phone companies use SS7, which defines a communications protocol and formats incoming and outgoing messages, similar to DSS1. However, SS7's signaling is much broader in scope and less specific than DSS1's signaling. DSS1 is specific to ISDN; SS7 handles the signaling needs of ISDN and older signaling systems.

SS7 is responsible for out-of-band signaling. Previously, the phone network claimed some of the user's bandwidth for call control. SS7 does this in a separate network, resulting in full 64-kbps user bandwidth and rapid call setups.

One important provision of SS7 is Common Channel Signaling (CCS). Keeping it simple, CCS makes it harder for mischievous users to harm the phone company's network. Also, it can improve service by making faster connections. Not all RBOCs have completed the conversion of their equipment to take advantage of CCS. Older equipment is still expecting signaling information in the same channel as voice, grabbing the eighth bit of each byte of voice data. As mentioned earlier, this is why your local access providers may offer only 56-kbps B channels in some areas. They continue to allocate one-eighth of their bandwidth to space for in-band signaling.

Aggregating Channels for Higher Bandwidth

Because ISDN is scalable in increments of 64 kbps, or increments of B channels, the telecommunications industry has developed a variety of combinations of B channels to deliver higher bandwidth to a single destination. Most users do not want to be limited to either BRI (two B channels) or PRI (21 B channels). ISDN service providers have responded with a range of B-channel aggregates between the two standard interfaces.

Unfortunately, clusters of B channels create a problem. For instance, two B-channel calls, originating from the same destination, may take very different paths while traveling across a public circuit-switched network. If data packets are divided or allocated equally across both B channels, a B channel's wild ride may cause the packets to arrive out of sequence at the other end.

Since customer premises equipment, such as an ISDN router, cannot control a B channel's paths, all aggregation techniques either attempt to negate or compensate for the B channel's erratic route. The techniques for aggregating B channels fall into two categories: dynamic and static. These are described in the following sections.

Dynamic Aggregation Solutions

With dynamic aggregation techniques, channels can be combined or split as needed, while maintaining the connections. There are three main techniques used as dynamic aggregation solutions:

- Dynamic bandwidth allocation
- The BONDING protocol
- Multilink Point-to-Point Protocol (MLPPP)

Let's see how each of these work.

Dynamic Bandwidth Allocation or "Rubber Bandwidth"

With ISDN, B channels can be combined for broader bandwidth to send larger files. Groups of B channels can be combined temporarily to support a full-color, full-motion videoconference or a large, high-speed file transfer. These same channels can be broken down and separated again for normal use, such as voice, fax, or e-mail.

The unique feature of combining B channels is referred to as *inverse multiplexing*. Inverse multiplexing (sending signals simultaneously) is the ability to add or subtract channels from another inverse multiplexed connection without terminating or losing the connection. This technique allows the bandwidth of a connection between two points to vary over time, accommodating changing traffic patterns or loads. It therefore provides a dynamic balance between economy and performance. This dynamic feature is often referred to as *dynamic bandwidth allocation*, or informally as *rubber bandwidth*.

> **NOTE**
>
> Inverse multiplexing can be defined as the sum of multiple individual digital channels across a network to create a single higher-rate information channel. For example, if six different independent 64-kbps data channels are established between points A and B, inverse multiplexing can be used to combine these channels to create a single 384-kbps (six B channels) pipe. If you find inverse multiplexing confusing, visualize traffic traveling on a freeway with several car pool lanes that can be opened and closed quickly, without closing the freeway, in response to the flow of traffic.

Dynamic bandwidth relies on some form of calculation to determine the load on one or more B channels. The load becomes a factor in the decision-making process, and the equipment (router) decides whether or not another B channel should be allocated. There are several proprietary algorithms schemes to share data across B channels, which makes interoperability difficult.

What You Can Do with Rubber Bandwidth

How could you use ISDN's dynamic bandwidth allocation feature? In videoconferencing, the quality of video motion (frames per second) increases as the bandwidth increases. For example, a videoconferencing session might use a 384-kbps (or three BRIs) inverse-multiplexed connection at the beginning of a teacher's lecture. The video camera may capture the teacher gesturing and moving side to side while writing on the board. The bandwidth (or channels) can be reduced when the video camera stops to focus for an extended period on a blackboard or visual aid. The relative lack of motion means less data is changing on the screen, and so less bandwidth is required.

In a LAN, an inverse multiplexed connection between two remote LANs might be established at 64 or 128 kbps, adequate for most simple interactive activities. However, if that connection becomes really busy (saturated) for a predetermined period of time (during a large file transfer, for example), the bandwidth of the connection can be automatically or manually increased to accommodate the increase in traffic (just like a car pool lane on a freeway). The increased bandwidth (or traffic lanes) would allow the file transfer to take place faster. When the file transfer is finished, the bandwidth would fall back to the original value.

The benefit of rubber bandwidth techniques is probably obvious at this point: Increased bandwidth available for all users and a cost savings in overall bandwidth by efficiently allocating bandwidth as needed.

BONDING

The competing proposal for providing this functionality is called BONDING, which approaches synchronization between multiple streams at the bit level. BONDING was designed for videoconferencing applications, and it usually requires additional hardware for the end system. Because BONDING is hardware-oriented, it is efficient. But it is also expensive and inflexible; once a pipe size is set, it cannot be changed until the session is finished.

When inverse multiplexers first began to appear, there was no standard inverse multiplexing protocol (or computer conversation etiquette). Therefore, each manufacturer developed its own proprietary protocol (remember the Beta and VHS formats for VCRs), which meant that both ends of an inverse-multiplexed connection required the same manufacturer's equipment. There was no interoperability between inverse multiplexers manufactured by different vendors.

Inverse-multiplexer manufacturers finally realized that without a multiplexer standard, the new data-communications technology might not reach its potential. Therefore, the Bandwidth ON Demand INteroperability Group (BONDING) consortium was formed by a number of inverse multiplexer manufacturers to develop an inverse multiplexing standard. The result was the creation of a BONDING specification, or protocol. (Most vendors also have their own proprietary methods, which usually add features and functions not present in BONDING.)

It is important to understand that BONDING is a specific protocol. It is designed for the circuit-switched aggregation of ISDN B channels to form a synchronous channel at the aggregate bit rate, as shown in Figure 3.3. In other words, BONDING puts a bunch of 64-kbps channels together and lets them think they are one big fat pipe.

FIGURE 3.3:

BONDING: An aggregate of two 64-kbps B channels, or two B channels joined together to look like one big data pipe of 128 kbps

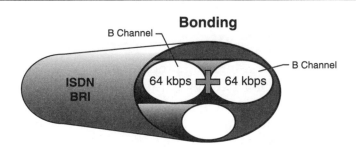

The BONDING protocol establishes how the transmission is handled:

- The channels to be used for the connection

- How the receiving inverse multiplexer will reorder the incoming data from the individual channels

- How the receiving inverse multiplexer will compensate for the relative delays between channels

The specification defines a way of calculating relative delays between multiple network channels and ordering data, so that what goes in one end comes out the other. Just imagine three of your friends leaving in separate cars after deciding on a destination and which freeway to use. After traveling in individual lanes, all three cars arrive in the same order as when they began the journey.

With BONDING, the router places data within a framing structure, which is then transmitted over multiple B channels. At the receiving end, the channels are phase-aligned and synchronized (using the framing structure) by the receiving router to re-create the original data stream. This framing and synchronization is processor-intensive and usually requires hardware assistance.

BONDING is most applicable for videoconferencing applications, where the delivery of information in the correct sequence is very important. It's also useful for remote LAN access, telecommuting, or whenever more bandwidth is needed.

MLPPP

The most common way to connect two computers over ISDN is through PPP. As explained earlier in the chapter, PPP is the standard protocol for transmission of IP (Internet Protocol) packets over serial lines, and it is the protocol typically used to access the Internet. For all its strengths, PPP has one inherent limitation when it comes to network deployment: it is designed to handle only one physical link at a time. MLPPP (or simply MP) does away with this restriction.

MLPPP is the new Internet Engineering Task Force (IETF) standard for B-channel aggregation. With MLPPP, routers and other access devices can combine multiple PPP links connected to various services into one logical data pipe.

The MLPPP standard describes how to split, recombine, and sequence datagrams across multiple B channels to create a single logical connection. It is written specifically for PPP. Major ISDN service providers and ISDN equipment vendors like 3Com, US Robotics, and Ascend are including support for the MLPPP standard in their products.

NOTE Technically, MLPPP is a higher-level, Data-Link protocol that sits between PPP and the Network protocol layer. It accommodates one or more PPP links, with each PPP link representing either a separate physical connection or a channel in a multichannel switched service, such as ISDN. See the section about the OSI reference model and ISDN standards later in the chapter for more information about communications model layers.

MLPPP differs from BONDING by specifying B-channel aggregation at the packet level rather than at the bit level. In addition, MLPPP does not influence the router's decision-making process concerning the allocation of extra B channels. Instead, it works to ensure that information packets arrive in the proper sequence, handling the re-sequencing or reassembly of incoming packets. MLPPP is the standard of choice among ISDN equipment vendors.

Using MLPPP, the two B channels can be combined for maximum throughput, as shown in Figure 3.4. Then, if you need another channel—to use the phone or send a fax, for example—the protocol dynamically allocates one channel to analog communications (voice, fax, or data). When the analog connection ends, MLPPP restores full bandwidth to the data connection.

FIGURE 3.4:

Two point-to-point connections coordinating the task of transferring a large file

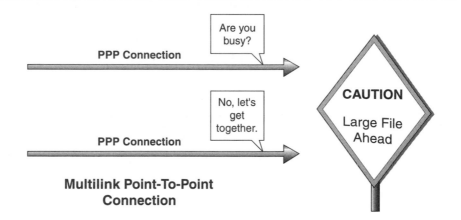

MLPPP takes advantage of the ability of switched digital services to open multiple virtual connections between devices to give users extra bandwidth as it's needed. MLPPP's ability to combine separate PPP links into one logical data pipe is one of the most important features of the protocol. Using a single phone connection, BRI provides two B channels for digital calls, each carrying up to 64 kbps of data or voice communication. Combining the two channels via MLPPP creates a 128-kbps data pipe between two Internet sites.

NOTE In order for you to use MLPPP, your Internet Service Provider (ISP), or your remote LAN access server at the office, will need to support PPP as well. If your Internet account is a SLIP account, you will need to have it changed to PPP. See the discussion of PPP and SLIP earlier in the chapter.

The basic version of MLPPP does not allow for dynamic control over individual B channels based on statistical bandwidth need. With basic MLPPP, you decide whether you want one or two B channels before you create a connection. The number of channels remains constant until the connection is closed. ISDN terminal adapters often differ in their implementation of MLPPP.

MP+, for Extended MLPPP, is Ascend's extension of basic MLPPP to allow individual control of B channels based on need. With this service, you usually open a connection with a single B channel. When 70 percent (user configurable) of your first B-channel's throughput is exhausted, the second B channel is dynamically (or automatically) added. Both B channels then combine to provide the 112- to 128-kbps throughput. As demand drops, and the second B channel is no longer needed (again, the user can configure the parameters), the channel is dropped and the connection returns to a one B-channel connection.

The chief advantage of MP+ is its efficient use of resources. Normally, ISDN connect time is metered or measured by B-channel connect time (B-channel minutes or hours). MP+ offers a significant cost savings by using the second B channel only when needed. Currently, several ISDN equipment vendors, such as Ascend and US Robotics, are implementing this protocol.

MLPPP, BONDING, and dynamic bandwidth allocation are dynamic aggregation solutions. Next, we'll look at static aggregation solutions and the H channel.

Static Aggregation Solutions

Static solutions do not offer the flexibility of allocating and deallocating channels as they are needed. There are two types of static aggregation solutions:

- Static dialing
- Multirate ISDN signaling and H channels

Static Dialing

With static dialing, a router decides to initiate the dialing of an entire group of B channels simultaneously. All the B channels share the load of that connection. This technique is no longer common because it does not provide any control for the effective use of bandwidth.

Multirate ISDN Signaling and H Channels

With Multirate ISDN signaling, the router sends a special call setup message to the local switch to identify the need for a call of predefined bandwidth. These calls are better known as *H-channel* calls. Table 3.2 shows the H channels that have been identified by the CCITT (now ITU), as specific rates above 64 kbps.

TABLE 3.2:   H-Channel Aggregation Rates

H Channel	Speed	Number of B Channels	Notes
H0	384 kbps	6	Most common
	1.544 Mbps		Uses 3 H0 channels
	2.480 Mbps		Uses 5 H0 channels
H1			
H11	1.563 Mbps	23	Equivalent to DS1 service
H12	1.920 Mbps	30	Known as European E1 service
H2			
H21	32.8 Mbps	512	
H22	44.2 Mbps	690	
H4	135.0 Mbps	2112	

H-channel calls are effective when you can determine the need for a predefined bandwidth in advance. A good example of this is videoconferencing applications, where 384 kbps of bandwidth is the normal requirement for good-quality video delivery. H-channel calls can also be effectively used as a backup for high-speed leased lines. H-channel calls are not effective where traffic flow is dynamic in nature, because bandwidth can go unused.

NOTE   Not all ISDN switch types currently support signaling that allows H-channel calls.

ISDN, LANs, and Internetworking

As its name implies, Integrated Services Digital Network was designed from the ground up as a digital service. Even voice calls are passed to the phone network in the form of a digital data stream. Because the phone network itself is digital, ISDN can interface to it directly. This is good news for computer networking. Like computer networks, ISDN is itself a digital network. However, ISDN is not limited by a network inside a building, campus, or even a city boundary. Network devices, such as bridges or routers, can output their digital signals directly without converting them to analog tones. End-to-end digital connectivity results in higher-speed connections and significantly reduced error rates. This makes WAN connections very efficient and reliable. ISDN's speed and digital attributes make it easy to connect to the LAN across the street or to an enterprise network half a continent away.

ISDN call charges are similar to conventional phone calls (metered by the minute), and when combined with the high data rates (bandwidth) available, make ISDN a good choice for LAN internetworking, particularly when communication is intermittent. Let's take a look at how ISDN compares with the other main technologies for linking LANs:

- **Leased lines:** These types of lines have fixed rental costs based on speed and distance and are usually installed for a minimum of one year. Costs are the same, whether there is any data being transmitted or not. This means that they are suitable for constant access, but can be prohibitively expensive for ad hoc communication.

- **X.25:** For X.25 networks, charges are based on line speed, call duration, and the volume of data sent. The data charge dominates the overall costs, making X.25 good for interactive applications but expensive for file transfer.

- **Dial-up:** Dial-up services are charged at the same rate as phone calls. They are limited by speed and are often poor quality, especially for international calls. The call setup time can also be quite long. Normally, dial-up services are used for short, infrequent communications.

- **ISDN:** For ISDN, charges are like those for a phone call, based on call duration, time of day, and distance. Because there is no volume data charge (for the number of bytes transmitted and received), it is most cost-effective for file transfer, a frequent LAN-to-LAN activity. Another major advantage of ISDN for LAN internetworking is that adding a new site is much easier and more cost-effective than with a leased-line solution.

NOTE   ISDN service can be a disadvantage in some client/server environments, particularly Novell NetWare networks. When a connection between a server and client is idle, packets are often sent between network nodes. In a LAN-only environment, these packets do not cause a problem; but when used across a WAN link, such as ISDN, the packets can significantly increase the cost of operating the network. *Spoofing* is a method of responding locally to these "keep-alive" packets, thereby optimizing WAN costs.

ISDN Internetworking Applications

Here are some examples of ISDN internetworking applications:

- **Direct connection into ISDN:** The ISDN router can be connected directly into the ISDN socket (NT1) supplied by the equipment vendor. Connection can also be made into an ISDN bus, where other ISDN equipment is sharing the ISDN line. Each piece of equipment (up to a maximum of eight pieces) will have a different extension number (subaddress) and all will contend for the two available B channels.

- **ISDN router through ISDN PBX:** This is a common application when you already have several ISDN lines in your office and want to share the use of outgoing lines, or when a PRI line connects directly into your PBX. ISDN supports direct dialing through a PBX, allowing incoming data calls to be sent directly to the router.

- **Bandwidth on demand:** The main advantage of ISDN is that bandwidth is available when required and only paid for when used. When data is being sent, the connection is open; when the line is idle, the connection is automatically closed. If access to a remote location is required for less than, say 20 minutes every business hour, ISDN is more cost-effective than a leased line. This is especially true for connecting overseas to major European countries.

- **Leased-line backup:** A common use of ISDN, particularly in the United States, is as a backup to leased lines. The leased line forms the primary link, and the ISDN connection is kept in reserve as a backup. If the primary link fails, the ISDN link may be used automatically. When the primary link becomes available again, the ISDN connection is terminated. ISDN bridges, described in the next section, are ideal for this application.

- **Remote PC access to the corporate LAN:** ISDN provides a cost-effective way to connect remote PC users with individual workstations to the corporate LAN. The PC requires an ISDN card and appropriate protocol support, such as an IP connection. The individual connection is made via an ISDN router. If the router supports PRI, up to 30 PCs can have simultaneous access to the LAN.

ISDN Internetworking Equipment

For connecting LANs with ISDN, you may need special terminal adapters, bridges, routers, or PC cards.

Terminal Adapters (TAs)

TAs are external devices that connect a conventional data interface, such as V.24, to an ISDN circuit, allowing non-ISDN equipment to interface with ISDN. Terminal adapters are widely used by internetworking manufacturers who do not have an approved native ISDN interface for their devices.

A disadvantage of this solution is that not all the information from the D channel passes through the TA, so non-ISDN equipment cannot take full advantage of ISDN facilities, such as calling-line identification.

NOTE Going by various names such as calling line ID, ANI (automatic number identification) and ICLID (for InComing Line IDentifier), this service provides the number of the calling party to the called party. ICLID is more than a convenience; it can be very useful for security purposes. The calling device's phone number can be used to validate network access for the caller.

ISDN Bridges

Because of its simplicity, bridging is one of the most popular ways of linking LANs. The big problem with ISDN bridging is controlling the bridge's use of the ISDN network.

Bridges are simple to set up and use because they will forward data, such as broadcasts, by default. With ISDN service, this means that calls will be made to send large amounts of unnecessary data. Over a period of time, this setup can prove very expensive. To avoid this, bridges can be configured to block broadcasts from specific addresses and to understand particular protocols. Unfortunately, the additional configuration causes the bridge to lose one of its major benefits—simplicity. However, bridges are ideal for backing up ISDN.

ISDN Routers

Routing is a far more effective way to apply ISDN to LAN internetworking, and the ISDN router approach is being adopted by all the major internetworking vendors. Using a router, data is sent over the ISDN network only when it is really needed. There are no unnecessary broadcast messages to transmit, so the bandwidth is used more efficiently than with bridges and the configuration is often simpler. Also, as a network grows, it's easier to add the router configuration than to set up more filters on bridges to block out all unnecessary traffic.

ISDN routers can dynamically allocate extra B channels for higher bandwidth applications. As the bandwidth increases, extra B channels can be opened, giving a much higher effective link speed. For example, Router 1 can aggregate extra B channels to Router 2 as required.

ISDN PC Cards

In many businesses today, there is an increasing need to connect home office users into the central LAN to enable resource and application sharing. ISDN PC cards allow individual workstations to connect to a central site that contains an ISDN router. This gives home users access to the resources of the entire corporate LAN on a dial-up basis.

ISDN and Your Home Network

The phone "network" inside of your house, shown in Figure 3.5, will be somewhat more complicated with ISDN than it is today, because it will be a true data network. This network is often called the *customer premises installation*, or CPI. In the future, your home network may consist of phones, computers, fax machines, video-conferencing, and an endless list of pie-in-the-sky applications.

FIGURE 3.5:

Your house's network

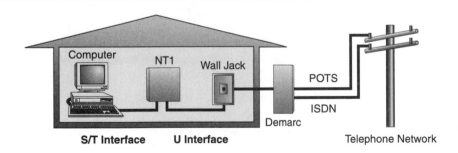

The following sections describe the parts of your communications system for ISDN service.

Your Network Terminating Device (NT1)

ISDN requires a network termination device, or NT1. The NT1 provides the physical and electrical termination of the twisted-pair wiring coming from the phone company's central office. The NT1 also converts the two-wire twisted-pair connection into an eight-wire distribution system within your premises. This provides a standard interface for ISDN terminal equipment. Any terminal equipment designed to meet the ISDN standard can plug into this interface by using a standard 8-pin RJ-45 connector.

The NT1 also provides remote diagnostic capabilities to allow central office personnel to perform centralized fault isolation. Some ISDN terminal adapters have the NT1 built in.

ISDN has two common interfaces, or connection points, called S/T and U. The S/T is a six-wire interface: one pair is for transmitted data, one pair for received data, and one pair for power. The S/T interface is on the customer (user) side of the NT1. The U interface is a two-wire interface to the phone company's central office. These interfaces are described in the following sections.

Products with built-in NT1s are much easier to connect. As Figure 3.6 shows, the U interface is directly connected to the device. This makes installation nearly as convenient as connecting an analog phone. In addition, some terminal adapters have an extra S/T port into which you can plug additional ISDN voice or data terminals.

FIGURE 3.6:

An external ISDN inte-
grated terminal adapter
(TA), with a built-in NT1
and power supply

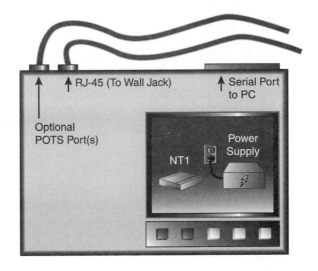

Fortunately, besides buying and plugging the device into the wall,
you have very little to do with the NT1. It functions as part of the
phone network. Most NT1s include LEDs (light-emitting diodes) that
provide limited information about the status of the ISDN connection.

Physical Connections to ISDN

Figure 3.7 shows the different types of functional devices that connect a customer site to ISDN services at the phone carrier central
office, as well as the ITU-TSS–defined interfaces that link them to
network services. Equipment that complies with these ITU-TSS–
defined interfaces is guaranteed to be compatible with both ISDN
and the other functional devices connecting to ISDN from the customer premises.

FIGURE 3.7:

ITU physical interface
reference points

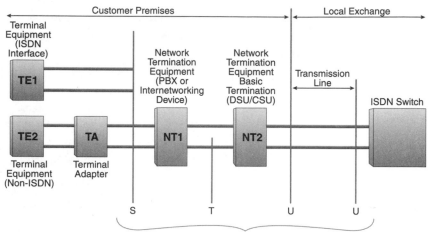

Terminal equipment (TE) consists of devices that use ISDN to transfer information, such as a computer, a telephone, a fax machine, or a videoconferencing machine. There are two types of terminal equipment:

- Devices with a built-in ISDN interface, known as TE1

- Devices without native ISDN support, known as TE2

Terminal adapters (TAs) translate signaling from non-ISDN TE2 devices into a format compatible with ISDN. TAs are usually stand-alone physical devices.

The S interface is an eight-wire interface, of which four wires are mandatory. It connects terminal equipment to a customer switching device, such as a PBX, for distances up to about 2000 feet. The S interface can act as a passive bus to support up to eight TE devices bridged on the same wiring. In this arrangement, each B channel is allocated to a specific TE for the duration of the call.

Devices that handle on-premises switching, multiplexing, or ISDN concentration, such as PBXs or switching hubs, qualify as NT2 devices. ISDN PRI can connect to the customer premises directly through an NT2 device; ISDN BRI requires a different type of termination. Like the S interface, the T interface is an eight-wire interface, with four wires required, that connects customer site NT2 switching equipment and the local loop termination (NT1). (T and S are identical, but their location is different; on a BRI, this interface it often referred to as S/T.)

As explained in the previous section, an NT1 is a device that physically connects the customer site to the phone company local loop. For PRI access, the NT1 is a CSU/DSU (channel service unit/data service unit) device; for BRI access, the device is simply called by its reference name, NT1. It provides at least a four-wire connection to the customer site and a two-wire connection to the network. In Europe, the NT1 is owned by the telecommunications carrier and considered part of the network. In North America, the NT1 is located on the customer premises.

The U interface is a two-wire interface to the local or long-distance phone company's central office. It can also connect terminal equipment to PBXs to support distances up to about 2 miles. The U interface is currently not supported outside North America.

Figure 3.8 shows a generic ISDN site. ISDN circuits are implemented by the provider as two-wire copper circuits from a central office within 18,000 feet of the user's office or home demarcation point.

NOTE The *demarcation point,* or *demarc,* is the point at which the customer's equipment and wiring ends and the phone company's equipment and wiring begins.

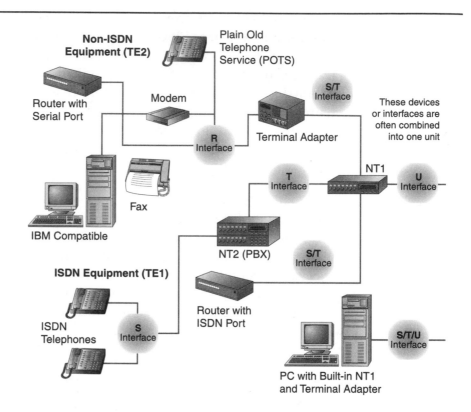

FIGURE 3.8:

ISDN equipment and reference interfaces

The RJ-11/RJ-45 ISDN interface, or two-wire U interface, at the demarcation point is then connected to an NT1 device. The U interface is defined by the ITU as the demarcation point of the two-wire ISDN subscriber loop.

The NT1 device is used to connect disparate equipment to the ISDN circuit at the U reference point. If a switched connection is needed, the network termination device (NT2) connects to the T reference point on the NT1 device and provides an S reference point to other ISDN-ready devices, such as telephones. A separate S/T reference point in the NT1 device will provide direct multipoint connection of ISDN devices if switching isn't necessary at the site. Two types of devices can connect to an S/T interface: TA and TE1.

In an ISDN implementation, TAs provide an R reference point, which lets non-ISDN analog and serial data terminal devices (such as modems, fax machines, POTS phones, and routers) connect to the NT1 device for ISDN service. Devices that require a TA are called TE2 devices. Multipoint TE1 connections can connect directly to the S/T interface on the NT1 device. For TE2 devices, the TA's physical interfaces can be RS-232 or V.35 serial interfaces. A combined NT1-NT2 device, which is more likely what you'll need, usually has an RJ-45 S/T interface to allow direct connection of four-wire ISDN devices, such as routers with built-in ISDN ports.

NOTE   To make life easier, the phone companies devised a lettering reference model to refer to the different points of an ISDN line. The U interface is the interface that comes into your house and extends from the wall outlet. This U interface is terminated at an NT1 which converts it to a T interface. This T interface is used by most terminal adapters (including some ISDN modems).

ISDN modems specify if they use a U interface (which means they have a built-in NT1) or if they use an S/T interface. More and more ISDN modems incorporate NT1s and connect directly to the U interface. The disadvantage of this is that you can use only the ISDN modem and you cannot chain more terminal adapters off a single ISDN line.

What this means to you is that once you have ISDN service, you need to know which ISDN interface your equipment expects: the S/T interface or the U interface.

The S/T Interface

The S/T interface defines the physical network inside your home or office. Wiring, connectors, and power are included in this interface.

The S/T interface utilizes an eight-wire interface (RJ-45), with four wires for signaling and four wires for power. In point-to-point applications, the S/T interface can connect equipment up to just over a half mile. When used in a passive bus configuration connecting up to eight terminals, it can span a distance of up to 1600 feet (this is called long passive; short passive connects up to 700 feet).

ISDN uses a phone jack that looks just like the standard phone jacks in use today, except that it is a bit wider. Instead of the older four-pin jacks (which used only two wires), ISDN uses an eight-pin jack (which uses four wires). The customer premises installation is based on a four-wire scheme: two wires for transmitting, and two for receiving (which means you'll probably need to rewire your house). Figure 3.9 shows the S/T interface in your house.

FIGURE 3.9:

S/T interface inside your house

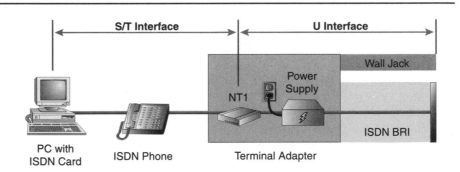

U-Connector (NT1)

The NT1 is the ISDN line terminator. Its function is to act as the bridge between the phone company wiring and the end-user wiring. A single NT1 can support multiple terminal adapters. If your equipment supports the S/T interface, you need to get an NT1 to convert between the U interface and the S/T interface. The NT1 has a jack for the U interface from the wall and one or more jacks for the S/T interface connection to the PC, other ISDN devices, analog devices, or an external power supply.

In the United States, the customer supplies the NT1, which includes the U interface connector. In Europe, the phone company, not the customer, supplies the network termination device.

The U Interface

U interfaces can drive a signal up to 18,000 feet. For distances longer than 18,000 feet from a phone company's central office, the phone company must install repeaters to boost the ISDN signal, as described in the next section.

The U interface has a single pair of wires with an RJ-11 connector that connects into the network. The RJ-11 connector is the type that the POTS uses to connect to the wall jack. Figure 3.10 illustrates this setup.

FIGURE 3.10:

The U interface with an RJ-11 connector

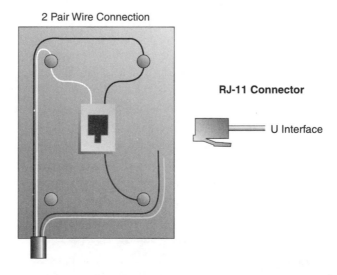

Some ISDN terminal adapters sold in North America connect directly to a U interface. If the PC is the only equipment to be connected to an ISDN line, this type of adapter is the easiest to install. In fact, the terminal adapter is surprisingly similar to an analog modem.

Both have external power supplies and a POTS port or two, providing an option to connect an analog phone, a fax machine, or an answering machine. Manufacturers may describe this feature as a "built-in NT1" or simply as a "U interface ISDN adapter."

Repeaters for ISDN

Because ISDN is a digital service, it must be configured precisely. To use the service, you must be within a given distance of the phone company equipment that serves you (typically 18,000 feet). Furthermore, there cannot be any other anomalies near the wiring that might interfere with the transmission. Even if the equipment at the central office is "ISDN ready," you may not be able to get ISDN due to line conditions or your distance from the central office.

To determine whether your particular wiring will support ISDN, the phone company will perform what is known as a *line qualification*. The technician will come to your location and place a load on the line. The success or failure of this test will determine if the phone company will need to install a repeater.

A *repeater* is a device that picks up a weak digital signal and then creates a strong copy that is retransmitted through the wire until it reaches the phone company's central switch. Because a repeater only needs to look for the presence or the lack of a signal (on or off, ones or zeros), the noise that usually "creeps" into the copper wire can be significantly reduced.

NOTE In an analog system, amplifiers are used to boost signal strength. Unfortunately, amplifiers tend to increase noise along with signal strength.

ISDN Cabling: Twisted-Pair

Two copper wires, each encased in its own color-coded insulation, are twisted together to form a twisted pair. Multiple twisted pairs are packaged in an outer sheath, or jacket, to form a twisted-pair cable.

Twisted-pair cable has been around for quite a while. In fact, early telephone signals were sent over a type of twisted-pair cable, and just about every building today still uses twisted-pair cable to carry telephone and other signals. However, signals have become more complex over the years, evolving from 1200 bps to over 100 Mbps. And there are many more sources of interference that might disrupt those signals today than there were at the turn of the century.

Coaxial cable and fiber-optic cable were developed to handle higher-bandwidth applications and to support emerging technologies. The good news is that ISDN uses existing twisted-pair wiring. This is a significant advantage for businesses in older buildings or offices that need to upgrade their communications system but can't afford to rewire. ISDN can be a low-cost alternative for digital communications when factors such as installation, existing infrastructure, and access are considered.

ISDN and Bearer Services

As explained at the beginning of the chapter, three types of switching service options are offered: CSV (circuit-switched voice), CSD (circuit-switched data), and PSD (packet-switched data). These are broad categories of *bearer services* that the phone companies can provide. Different bearer services provide different types of guarantees about the reliability and synchronization of the data. There are currently ten different bearer services for circuit-switching mode and three services for packet-switching mode.

WARNING In certain areas, phone companies aren't prepared to dynamically provide whatever service you need. When you order ISDN, you will be asked what type of information the ISDN channels are going to transmit. Make sure that you know what type of B-channel configuration is being provisioned. You could spend hours trying to get your ISDN equipment to work, only to find that a channel was provisioned with only voice or data.

These bearer services are defined in terms of a number of attributes, which include mode (circuit or packet), structure (bit-stream or octet-stream), transfer rate (such as 64 kbps), transfer capability (basically, the content, for instance speech, 7 KHz audio, video, or unrestricted), and several other attributes that specify protocols and other information.

Which Switch Is Which?

With pure ISDN, switching shouldn't be your concern; after all, the central office switch is the phone company's headache. However, because ISDN is an evolving technology, the local phone companies are in the middle of upgrading their equipment, so we need to mention switches.

The magic behind ISDN is the intelligent central office switch. These devices are really large mainframe computers with many input/output ports. There are a number of switch vendors offering ISDN central office and PBX switches. The big three switches in North America are the AT&T 5ESS, Northern Telecom's DMS100, and Siemens EWSD.

It is important to know which switch is being used to supply your ISDN service. Some ISDN devices only work with a particular switch. Even devices which work with a variety of ISDN switches can experience compatibility problems. Be sure to ask your service provider which type of ISDN switch serves your location. In most cases, incompatibility will not be a problem, because of the adoption of the National ISDN standard, described later in the chapter.

You'll also need to know which type of switching service you need. If you are using CSV, any type of switch will do—even the old analog switches (yes, they still have a few left out there). Digital data may even be shared with other channels in the moments when there is silence on your phone line. Just imagine, every time you stop to take a breath during your phone conversation, the phone company can "sneak" some digital data through the network.

For CSD, the phone company can't get away with older equipment. Digital data calls must be routed onto pieces of equipment that will supply dependable, full-time data channels.

Fortunately, in many cases, the phone company can provision your ISDN B channels to handle both CSV and CSD, at no extra charge. Typically, you can get both B channels for data, or one for voice and the other for data, or both channels for data and voice. If available, order both B channels to carry voice and data. However, in some areas, you may still need to tell the phone company how you are going to use your B channels, so you are provisioned with the appropriate CSV or CSD channel.

NOTE   Some terminal adapters include a feature that allows you to use a CSV bearer service to carry data (perhaps because it is cheaper, or possibly CSV is all that is available), which is called Data Over Speech Bearer Service (DOSBS). This works by providing additional end-to-end data guarantees that can't be relied upon from the speech bearer service.

What Is a SPID?

A regular phone line has only one phone number attached to it and never can be shared at the same time by two different applications (such as two different phone calls). ISDN lines are different. A single ISDN line can be used for many different applications at the same time, such as a phone, computer, fax, and other applications.

To keep track of all these different applications, ISDN uses a SPID (Service Profile Identifier) to identify each application. One (or more) SPID is assigned to each application on the ISDN line. Therefore, a major difference between an ISDN line and a POTS line is that a POTS line has only one directory number, but an ISDN line may have many.

The SPID is used to identify the ISDN device to the phone network, much as an Ethernet address uniquely identifies a network interface card. Without a SPID, which looks like a phone number with extra digits, ISDN devices won't work on most lines. Some ISDN switches require only one SPID per line. Others require one for each channel. The phone service provider in your area should be able to tell you what your SPID is and how many SPIDs are required.

Currently SPIDs are used only for circuit-switched service (as opposed to packet-switched). They are most commonly implemented by ISDN equipment used in North America. The Euro-ISDN standard used in Europe and other countries does not require SPIDs.

When a new subscriber is added, the phone company personnel allocate a SPID, just as they allocate a directory number. In many cases, the SPID number is identical to the (full ten-digit) directory number. In other cases, it may be the directory number concatenated with various other strings of digits, such as digits 0100 or 0010, 1 or 2 (indicating the first or second B channel on a non-Centrex line), or 100 or 200 (same idea but on a Centrex line), or some other, seemingly arbitrary, string.

Subscribers need to configure the SPID into their terminal (that is, the computer or phone, not the NT1 or NT2) before they will be able to connect to the central office switch.

When the subscriber plugs an ISDN device into the line, some initialization takes place, establishing the basic transport mechanism. However, if the subscriber has not configured the given SPID into the ISDN device, the device will not perform the next level of initialization, and the subscriber will not be able to make calls. This is, unfortunately, how many subscribers discover they need a SPID.

Once the SPID is configured, the terminals go through an initialization/identification state, which has the terminal send the SPID to the network in a message. Thereafter, the SPID is not sent to the switch again.

There are some standards that call for a default Service Profile, and a terminal doesn't need to provide a SPID to become active. Without the SPID, however, the switch has no way of knowing which terminal is which on the interface. For multiple terminals, an incoming call would be offered to the first terminal that responded rather than to a specific terminal.

What Is a TEI?

The TEI (Terminal Endpoint Identifier) is used with the SPID to identify individual devices on the ISDN line. Unlike the SPID, which must be set or input by the user while configuring ISDN equipment, the TEI is usually set to "auto," allowing the phone company's central switch to assign the TEI for the device.

TEIs identify the terminal for a particular interface (line). TEIs will be unique on an interface; SPIDs will be unique on the whole switch and tend to be derived from the primary directory number of the subscriber. TEIs are dynamic (different each time the terminal is plugged into the switch), but SPIDs are not. Following the initialization sequence, the one-to-one correspondence is established. TEIs are usually not visible to the ISDN user, so they are not as well-known as SPIDs.

What Is a BC?

The attributes of the bearer service are encoded into a bearer code, or BC, that is sent every time a new connection is being set up. In theory, this allows the switches to dynamically choose from a variety of different switching-path techniques depending on requirements. In practice, the SPID is used to determine what services are needed for switching.

SPIDs are generally used in only the United States and Canada. Using the SPID greatly simplifies things for the phone companies. However, the BC will not be completely ignored; there are certain bearer services that will be unavailable on your B channels, based on how they are configured.

It is important to note that the BC is sent to the switch every time a connection is established. The SPID is sent to the switch only when you physically attach your equipment to your phone line. At this time, the switch gives your device a TEI, which is used from then on to identify all connection requests from that piece of equipment. This allows the switch to look at the TEI and BC, determine the SPID, and see if the BC and the SPID match up.

ISDN Ordering Codes

An ISDN ordering code is a preestablished set of network services associated with specific customer equipment for a specific application. ISDN ordering codes have been adopted by many customer premise equipment (CPE) vendors in order to alleviate what is currently a major source of confusion: the process of identifying the necessary equipment and ordering the appropriate ISDN service from a service provider.

The codes are based on network configurations already used by thousands of ISDN customers across the country. The use of these codes allows for maximum compatibility with minimum customer knowledge of line provisioning and other complexities. All you need to do is give your service provider the ordering code specified with your equipment purchase.

ISDN Line Types and Configurations

ISDN lines are provisioned based on the characteristics of the local loop. As mentioned earlier in the chapter, before the local company attempts to install the ISDN line, it will perform a line qualification on the cable pairs. This involves a series of measurements to ensure that the cable pairs meet the system requirements.

Lines for T and U Interfaces and Z Cards

All ISDN lines are equipped with an individual line card in the 5ESS switch. ISDN users who are in proximity to the local ISDN service office may be provisioned on a T interface line. A T interface provides a direct ISDN connection to the 5ESS switch. This is a four-wire connection. The length restriction for a T interface line is approximately 3300 feet of 22-gauge wire (or more accurately, 6 dB loss at 96 KHz). A T interface delivers the ISDN signal to the terminal without the need for an NT1 unit.

The U interface line requires a different configuration due the length of the local loop to the user's terminal. First, the U line requires a different line card in the 5ESS switch. This is a two-wire interface. Two types of U line cards can work in the 5ESS switch: the AMI line card, which is proprietary and being phased out, and the newer 2B1Q ANSI-compatible line card. The length restriction on the 2B1Q U line is 27,000 feet of 22-gauge wire (or 42 dB loss at 40 KHz). The U interface also needs to have an NT1 unit at the user's location. The purpose of the NT1 is to provide digital loopback capability and to convert the 2B1Q line code to bipolar signaling for delivery to the ISDN terminal.

If you want to operate analog lines from ISDN digital equipment, analog lines can be provisioned by using a Z card in the ISDN line unit. However, the use of Z cards should be kept to a minimum, because the Z card forces the 5ESS digital switch to perform double duty to make the line work. The Z card was designed for situations where an analog line may be required in a location served by predominantly ISDN lines. Z cards are generally used for phones in areas that are difficult to reach or for pay phones.

NOTE All ISDN terminals require a power connection (although the power connection may not be evident because the power units are generally located in wiring closets). The power is used for the terminal codec and also to provide all of the signaling tones. Unlike an analog phone, where a ringing generator (105 volts) is placed on the line to ring the phone, in ISDN service, an "Alerting" call control message is relayed to the terminal. The Alerting message tells the ISDN terminal to play the alerting tones. In fact, the dial tone originates in the ISDN terminal.

Point-to-Point versus Multipoint Lines

ISDN lines may be configured as point-to-point lines or as multipoint lines. The difference refers to the wiring configuration and to the way the line is defined by the software in the 5ESS switch.

ISDN lines wired as *point-to-point lines* are simply one ISDN terminal assigned to an ISDN line card. This one terminal has full access to the two 64-kbps B channels and the one 16-kbps D channel. The user on this DSL (digital subscriber line) can use the channels independent of each other and simultaneously. The phone number and all of the available options and features assigned to the line are logically linked to the DSL location.

In a *multipoint line* configuration, more than one ISDN user is assigned to a DSL. Users share the two B channels and one D channel. When more than one terminal is assigned to a DSL, the system must identify which terminal to signal for incoming connection requests and which terminal wishes to initiate an outward connection. To aid in this process, two attributes are assigned to the terminals: the SPID and the TEI. (Actually a TEI is assigned to all terminals, including point-to-point lines; for point-to-point lines, the TE1 is a default value and not used in the identification process.)

For example, with a multipoint line, when an incoming call is received at a phone number, the 5ESS switch will deliver the call over one of the available B channels. When a second call is received on that same phone number, the second call appearance will flash and the caller's line ID will appear in the Line ID LED display. If the user places the first call on hold to answer the second one, the first call appearance will flash to indicate its on-hold status. The first call, now on hold, will be held at the switch line card in the 5ESS, and the second call will use the same B channel. It is not possible to use both B channels for voice calls. Even for a conference call, you don't use the second B channel, because the conference-call bridging is done in the 5ESS switch and the resulting conversation is delivered on the one B channel. If the ISDN service is used primarily for voice-only applications, there will be unused capacity on the ISDN DSLs. It is much more cost-effective to use multipoint lines in place of point-to-point lines for voice applications.

ISDN Standards

In 1991, Bell Communication Research Corporation (Bellcore) issued a technical specification developed to standardize ISDN services offered by the seven RBOCs. Called National ISDN, the specification serves as a recommendation to other communications companies developing product and services in the United States. The goal of National ISDN is to make ISDN a seamless service that is easily accessible nationwide.

> **NOTE**
> In Europe, most ISDN services conform to the Euro-ISDN (European ISDN Standard), which was introduced in 1988 by the European Telecommunications Standards Institute (ETSI). See Chapter 4 for more information about ISDN standards and services outside the United States.

National ISDN Specifications

There are three National ISDN (NI) specifications: NI-1, NI-2, and NI-3, each building on the previous one. ISDN devices that specify NI-x compatibility tend to have a high degree of compatibility with various ISDN central office switches.

Most ISDN equipment vendors now market equipment designed to meet the NI-1 specification, and carriers have upgraded their central office switches with software based on N1-1. Some phone companies have also incorporated N1-2. NI-3 may be available from vendors by next year. Table 3.3 summarizes the three phases of the NI specifications.

> **NOTE**
> Even though NI-2 attempts to standardize PRI, most carriers are still offering proprietary versions of PRI, because they feel it has more features.

TABLE 3.3: National ISDN Standard Implementation Phases

Phase	Description
NI-1	Provides standardization for BRI.
NI-2	Enhances NI-1 by establishing service uniformity.
	Defines operations and maintenance improvements.
	Provides a signaling standard for PRI.

TABLE 3.3: National ISDN Standard Implementation Phases (continued)

Phase	Description
NI-3	Enhances NI-1 and NI-2.
	Adds music on hold and calling name delivery.
	Enhances operations to support billing capabilities.
	Adds interfaces between ISDN and other network services, such as frame relay and personal communications system (PCs).

What Are Custom and Standard ISDN?

Several interexchange carriers, including AT&T, MCI, and Sprint, do not adhere to the National ISDN recommendations exclusively. Since NI-1 specifications do not include a number of features—and thus equipment connections—that these companies believe are important, they currently offer a separate version of what is generically called "Custom ISDN." They sometimes call the services that conform to the National ISDN specifications "Standard ISDN."

As an ISDN service user, you do not need to be concerned about the differences between Custom and National ISDN. The differences mainly have to do with the internal procedures and the specific definitions of various services. In many cases, telephones, terminals, and other devices designed as Custom ISDN services work with the ISDN systems offered by regional and other carriers. If you need to reconfigure components that work with one type of ISDN to work with another type, you can usually make this change through a network configuration utility.

ISDN Standards Development and Product Competition

The evolution of the National ISDN and Custom ISDN standards began with the early days of ISDN, when the technology was being developed along proprietary lines. In the United States, the two competing camps were AT&T with its 5ESS switch and Northern Telecom with the DMS100. Both products were rolled out at approximately the same time and had similar features and options, but they were not compatible for data connections. Although the voice services were somewhat similar, it was not possible to connect a data call between the AT&T and Northern Telecom equipment. The introduction was also confusing to potential customers. Each switch offered similar features, but the terminology used to describe the services and the way that the features worked were much different.

As a result of the compatibility issues and customer confusion, ISDN in the late 1980s was an "island" application. Those who purchased the service did so to use the ISDN capabilities on a Centrex or an intraoffice basis. The promise was that interconnection was in the near-term future. Escalating the situation was the environment of deregulation of the phone industry. There was no clear direction on where the network was headed, who the dominant players would be, and how to bridge the gaps of interoperability.

Part of the deregulation of the telecommunication industry was the establishment of Bellcore. With the separation of AT&T from management of the local phone network, the newly formed regional operating companies needed a research organization to take the place of AT&T's Bell Labs. The company formed was Bell Communication Research, or Bellcore. It was made up of former Bell Labs personnel working on the local issues of the telecommunications network. As one of its initial tasks, Bellcore published a series of ISDN standards documents describing ISDN services in generic terms, removing the vendor-specifics from ISDN service. These standards provide a platform upon which a number of equipment manufacturers can produce and market components and services on the public network.

When the standards documents were finally published, each switch manufacturer (AT&T and Northern Telecom) had a somewhat extensive installed base of ISDN customers. All of these applications were proprietary and, therefore, did not meet the new ISDN standards. The companies now faced the prospect of changing their products to meet the new standards while maintaining support for existing ISDN networks. The companies now refer to two types of ISDN: Custom ISDN and Standard ISDN. Simply stated, Custom ISDN refers to components and service that use procedures proprietary to a specific manufacturer. National ISDN or Standard ISDN are those components and services that conform to the Bellcore ISDN standards.

The acceptance of the National ISDN standard has motivated the communications industry to begin support of ISDN in earnest. Because most ISDN equipment conforms to this standard, the phone companies now know how to set up their ISDN-capable switching services.

ISDN Standards and the OSI Reference Model

The standards used to define ISDN (both Custom and National) make use of the OSI (Open Systems Interconnection) reference model. This is a generic, seven-layer model developed by the ISO (International Standardization Organization) to define how devices can be connected for communications purposes. The layers are, from the bottom up, the Physical, Data-Link, Network, Transport, Session, Presentation, and Application layers. The modular approach allows developers to focus on a specific level of the model, knowing that the lower and higher layers will handle the requirements at their respective levels. For ISDN, we are concerned with the first three layers of the OSI reference model.

ISDN and the Physical Layer

The Physical layer of the OSI model defines the physical properties of an ISDN circuit. For example, it addresses these types of questions:

- What is the connector type?

- How many leads are in a cable?

- What is on pin 1, pin 2…?

- What defines a pulse?

- Is it 1 volt or a range from .75 to 1.18 volts?

All of these parameters make up the Physical layer specifications of the ISDN line. In order to specify the electrical standard, everyone involved must agree on where to measure these requirements. This agreement has resulted in the definition of user/network interfaces.

Physical Layer Interface Definitions

The interfaces for an ISDN Basic Rate Interface are as follows:

- **U interface:** Particular to the United States, the U interface is necessary due to the longer local loop lengths common in this country. When the local loop length exceeds approximately 3300 feet (1 kilometer) of 22-gauge wire, the strength of the signal must be increased. The older AT&T proprietary version was to use Alternate Mark Inversion (AMI) line code. The current standard is the American National Standards Institute (ANSI) 2B1Q line code.

- **NT1 unit:** This is the component on the user's side of an ISDN U line. The purpose of the NT1 is to convert the incoming line code from AMI or 2B1Q to bipolar. The NT1 also provides the demarcation between the public phone company network and the customer premises equipment. There is digital loop-back capability in the unit for testing purposes. The NT1 device is owned by the customer. The output of the NT1 is the T interface.

- **NT2 unit:** This is a device that provides multiple ISDN interfaces on an ISDN line. The NT2 may be as simple as a bridging device connected to an NT1 unit, or it may be as complicated as a PBX. The interface to the NT2 device is the T interface.

- **TE1:** This terminal equipment is provided with an ISDN interface capability, and it will receive and process bipolar digital signals from the network. This equipment may be a computer, fax machine, or a PCMIA. The interface to this component is the S interface. Basically, the S and T interface are electrically identical. The equipment may be served directly from the office if the loop length is short, or it may be a U line equipped with an NT1, which thus provides the S/T interface.

- **TA:** This is the terminal adapter, or the ISDN device that adapts non-ISDN equipment to work on an ISDN line. The purpose of the TA is to convert the bipolar signaling from the public network to the unipolar signaling used by computers. The most prevalent use for a terminal adapter is to connect a computer to an ISDN line.

- **TE2:** This terminal equipment does not have an ISDN interface. It needs an ISDN terminal adapter to function on an ISDN line.

NOTE
The technical specifics of the Physical layer specifications for ISDN are contained in the CCITT (ITU-T) documents. The specification for Basic Rate Interface is document I.430; for the Primary Rate Interface, it is document I.431.

ISDN Connector Types

The Physical layer also defines the type of connectors used. ISDN primarily uses three types of connectors:

- RJ-11, which is a four-conductor phone cord that can be used to connect an ISDN U line to the line side of a power source

- RJ-12, which is a six-conductor ISDN cord that can be used to connect from a power supply to the line side of the NT1 unit

- RJ-45, which is the eight-conductor cord used for ISDN

ISDN and the Data-Link Layer

In the OSI reference model, the Data-Link layer provides for procedures established to maintain communications between two network components. In the case of ISDN, the two components are the ISDN terminal and the ISDN switch. This means that ISDN terminals are in constant communication with the ISDN switch. An example of this is when an ISDN terminal is plugged in, a red LED illuminates. This LED is an indicator that layer 2 is established with the ISDN switch.

NOTE You'll notice that when a terminal is plugged in, the LED does not light up immediately; it may take a minute or so. The delay is the time it takes for the terminal and switch to negotiate and reach agreement on the means of communication. Once there is agreement on the procedures, a layer 2 connection is established, and the red LED will come on.

The procedures define the protocol to be used, which is Q.921. The protocol specification defines the frame structure of the data packets, the procedure elements, the format of the fields in a frame, and the procedures themselves, known as Link Access Procedures-D Channel (LAPD). The LAPD procedures describe flags, sequence control, flow control, retransmission, and similar items.

ISDN and the Network Layer

The Q.931 defines the layer 3, or Network layer, specifications for ISDN. The Network layer provides procedures to make end-to-end connections on the network. Remember that the ISDN connections established are for circuit-switched voice (CSV) and circuit-switched data (CSD) connections. Connections that involve X.25 packets conform to the X.25 protocol and different connection procedures. Figure 3.11 shows the Network layer and D-channel protocols.

FIGURE 3.11:

D-channel protocols and the Network layer

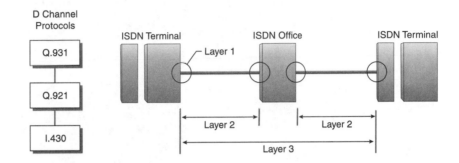

Q.931 procedures use packetized messages to initiate, monitor, and release circuit-switched connections. Call procedure messages are relayed between the ISDN terminal and the ISDN switch as the "information" part of the Q.921 information frame. Each call control message has a number of mandatory information and optional elements. What is mandatory and what is optional depend on the circumstances of the message. If, however, any mandatory information elements are missing, the message will be rejected and the call abandoned.

The X.25 protocol defines the Data-Link and Network layer (layers 2 and 3) specifications for packet connections. This protocol defines packet frame and connection procedures to establish and maintain a data link between two network components. An example of this type of packet connection is an ISDN terminal connection to a 5ESS switch, which may be used to establish a D-channel X.25 connection to a modem pool. The packet frame follows the same format as the HDLC frame in Q.921 protocol specifications. The procedure is defined as Link Access Procedure-Balanced (LAPB).

International ISDN Connections

- Why ISDN is popular in other countries

- The development of a European standard for ISDN

- ISDN equipment in other countries

- A country-by-country assessment

- Lessons from international ISDN implementations

- ISDN connections from the United States to other countries

If you do business or have friends or family members in other countries, you will be interested in ISDN connections outside the United States. Much has been written about the rapid embrace of ISDN in Europe, and many of us are wondering when the adoption rates for ISDN in North America will mirror those in Western Europe. ISDN connections are very widespread in most countries in the European Union (EU), and many more are coming online quickly.

ISDN is the most rapidly selling telecommunications service on the continent. Germany by far leads Europe in the amount of connections, with over 1 million Basic Rate ISDN lines and over 2 million B channel terminations so far. This is more installed lines than the year 2000 projections for any Bell company or RBOC (NYNEX and Pacific Bell have announced the target of one million installed lines by the year 2000.)

NOTE   The worldwide growth of ISDN revenue in 1995 was over 150 percent and projected to be above 100 percent again in 1996. Its growth during these years is second only to frame relay's growth.

The Global ISDN Experience

The popularity of the ISDN technology in many international markets is one of the reasons for the increasing demand for ISDN in the United States. As early as the late 1980s, Japan and Germany were making great strides in the conversion of many basic business services to ISDN. Now many countries are rushing to add ISDN services in their public networks. This is happening as they realize that telecommunications is a requirement for their basic economy and business community. ISDN came in at a time that this new investment was being considered for many of these countries.

Why There and Not Here?

In many countries, the conversion to ISDN has been heavily funded by the government-owned public telephone ministries (or PTTs, for Post, Telephone, and Telegraph agencies). It was easy for the phone company, with its monopoly on communications services, to make the decision to create a country-wide network based on ISDN digital technology. It also was easy to standardize; since only one company was involved, a common platform and standard interface were ensured. The public telephone companies in each country could work toward providing standard ISDN services, without any interoperability issues in their implementations.

These countries were able to provide a more efficient, digital service which primarily serviced the business community for basic voice applications. Now they are beginning to focus on data applications using ISDN, with particular interest in videoconferencing systems.

The rise in PC ownership in other countries has recently driven the need for a more reliable digital network. As sales in the United States start to slow down, PC manufacturers are targeting the international market. Most of the double-digit PC growth is expected to happen in other countries.

For several reasons, these countries have the opportunity to leapfrog technology and deploy ISDN as a basic infrastructure. In some countries, ISDN services can be more available and affordable than regular analog dial-tone. Some countries have had the luxury of introducing this new technology after it had already been tested and proven elsewhere.

In addition, some of these countries do not have a huge legacy analog dial network or a great number of modem users to address as does the United States. These countries are free to start fresh with the most universal digital technology that the mass market can use, without worrying about the legacy users or their stranded investment in analog switching equipment and support. They can install newer equipment that does not need to be compatible with any existing switch or the equipment that customer has already installed.

Currently, most ISDN applications in Europe are voice-oriented. The European market has been slightly behind the United States in developing the more complex data applications of ISDN, but some countries are catching up. One indication of this advancement is the recent announcement by Prodigy that it is increasing its services in Europe. Many vendors are also seeing this movement reflected in the increased demand for internetworking hardware, remote-access products, and business applications that work with ISDN.

ISDN is being widely deployed throughout Europe, and with a higher concentration in both metropolitan and rural locations than experienced in the United States. Germany and the UK are the countries making the most immediate plans to launch ISDN into the mass market. Italy and Spain have the lowest percentage of deployment, but ISDN availability is increasing rapidly in those countries. By the end of 1996, nearly 95 percent of Europe will have ISDN access, especially in western Europe.

The Euro-ISDN Standard

In the 1980s, when the ITU (International Telecommunications Union) began to specify the standards that would lead to the development of ISDN, the idea of digitizing public-switched telephone networks (PSTNs) was implemented in a variety of manners in different countries, similar to the experience across the United States. This made it difficult to communicate across borders and slowed down the deployment of ISDN.

In 1989, 26 European network service providers signed a Memorandum of Understanding with regard to the deployment of a common standard by 1993. European standards were drawn up by the European Telecommunications Standards Institute (ETSI), established in 1988. Once a country or PTT agrees to the Memorandum of Understanding, the ETSI standards require mandatory implementation. The term *Euro-ISDN* is used to describe this common ISDN implementation and agreement, which has increased the interoperability of ISDN services in Europe.

NOTE

The ETSI was created to take over the telecommunications standards work that previously was done by the CEPT (Conference Europeane des Postes et Telecommunications). The CEPT members were exclusively PTTs. ETSI has opened its membership to manufacturers and users.

In December 1993, the IMIMG (ISDN Memorandum of Understanding Implementation and Management Group) together with 66 equipment and solution providers, and with the support of the EU, launched Euro-ISDN in a three-day event called Eurie '93. This event spanned 75 cities in Europe and hosted over 30,000 people. Since then, ISDN growth has picked up dramatically. By the end of 1996, more than 7 million B channels (both Basic and Primary Rate Access) will have been deployed throughout the continent. By the end of the century, more than 20 million B channels are expected to be in use.

The acceptance of the Euro-ISDN standard makes it much easier for manufacturers to produce ISDN-compatible equipment. And because of this European standardization process, ISDN equipment manufacturers now need to test their equipment for compliance to the Euro-ISDN standard only once. After the equipment is tested and complies, then it is accepted in all the countries that have agreed to the Euro-ISDN standard, without further testing. This has greatly reduced the testing and approval of ISDN equipment in Europe and in the other countries that have agreed to the standard. The equipment manufacturers pass on their savings to the users by reducing the cost of ISDN equipment.

NOTE One interesting aspect of the Euro-ISDN standard is that it eliminated the need for a SPID (Service Profile Identifier). This approach is something that is currently being debated in the carrier and vendor forums on ISDN today. This may lead to the United States adopting the same type of standard, with a simplified, standard or embedded ISDN service.

IMIMG continues to gain members from Europe, the Far East, and Africa, and has grown to more than 35 network service provider members. In the November 1995 meeting of the IMIMG in Portugal, Poland signed the Memorandum of Understanding for ISDN, and joined South Africa, Australia, Malaysia, and Indonesia as recent signatories. The IMIMG has decided on further development of the Memorandum of Understanding to include plans for deployment in 1997 and continuing to expand beyond the European countries.

The EU places a lot of emphasis on telecommunications standards, and the governments have worked closely with the service providers to develop the ISDN infrastructure. The goals of the EU for ISDN are:

- To promote the deployment of Euro-ISDN, a common, vendor-neutral, standards-based ISDN platform

- To gain more worldwide adoption of ISDN based on ETSI (European Telecommunications Standards Institute) standards

In addition, the European Public Telecommunications Network Operators' (ETNO) Association, which represents 33 network service providers from 24 countries on the Europe continent, has sponsored ISDN and the Euro-ISDN standardization since 1993. The members of this association feel that ISDN is necessary for European (and non-European) countries to stay competitive and keep up with information-exchange technology. The ETNO looks to the IMIMG to develop and maintain the implementation of ISDN at the working level and expand beyond just European countries.

TIP You can get background information and full details on ETNO positions, including updates, through the Internet. See the association's home page at www.interpac.be/webspace/etno.

International connections are further being helped by standardized international ISDN links, which are being upgraded to the agreed-upon ISUP standard. ISUP stands for the ISDN User Part, which is the message type used in the network signaling for ISDN. Most European countries can provide connectivity to more than 40 international destinations. What this means is that the international connections of the United States and other nations will have standard interconnection protocols for ISDN calls and signaling, making these connections transparent to the users of ISDN.

ISDN Equipment Development in Other Countries

In Europe and many other countries, the network terminating device (NT1) is part of the telephone network that is supplied for the end users. This is one major reason why the terminal equipment for each country may have a different interface and software than equipment in the United States.

Because of the difference in interfaces and the lack of widespread ISDN use, the makers of ISDN equipment found it difficult to enter the United States market. However, now that ISDN is becoming more common in the United States, European and Japanese equipment manufacturers are finding that it may be profitable to turn their attention here. Without investing too much, many of the foreign companies can extend their channels or modify their existing ISDN equipment to work in the United States.

Fujitsu of Japan, Siemens of Germany, Sagem of France, and many others have already entered the market and started to see their ISDN equipment gain increased acceptance. Now they are not only making headway into the market, but some of these companies are being purchased for their ISDN expertise and technology (for example, Fivemere from the UK was purchased by Network Express). They are also spending some research-and-development dollars to develop more focused data applications for their equipment in the United States.

Another example of the newer European ISDN hardware coming into the United States is the product trials of Verifone company's credit card processing terminal with an ISDN interface. (Verifone is a United States-based company that has a contract with a German company to build these terminals.) This company is creating a new model to include this Tranz330-like model with new software and networking features. The ISDN interface will increase the reliability of transactions and make them much faster than they are with analog phone service.

NOTE

Some of the early Apple Computer ISDN interface hardware for Apple's initial version of ISDN-capable Geoport was developed by SAT-Sagem of France. This was Apple's strategy to forge a quicker entry into the market for ISDN-compatible customer premises equipment for Apple products. In most cases, any Apple ISDN terminal adapters will be devices external to the desktop unit. However, as Apple works to integrate many computer telephony features in its audio/video desktops, perhaps by early 1997 Apple will have integrated ISDN terminal equipment into desktop units.

Along with the more established companies, many new companies have entered the business of selling ISDN equipment in the international market overall; these companies are also considering expanding their reach into the United States. Similarly, many United States manufacturers are going abroad with their products.

Again, the key difference in many of these countries is that they consider the NT1, or network terminating device, to be a network element. In the United States, the FCC has ruled that the NT1 is CPE (customer premises equipment) and not part of the network.

Global '95 and '96 for Worldwide ISDN

In November 1995, Western Connect Ltd. and Bellcore coordinated an event that demonstrated the worldwide nature of the ISDN service. Live ISDN applications worked on a global basis in over fifty countries spanning five continents.

A promotional event of this scale for marketing a network service has never been attempted before. Many national and international companies sponsored and participated in this event. The promotion of Global '95 itself was lead by IGICOM. Seven major publications (*Communications*

continued on next page

News, Communications Week, Data Communications International, Electrotecnia Revista International, European Communications, Network World, and *Telecommunications International)* contributed to the Global 95 event by ensuring its worldwide promotion and allowing information sharing after the event itself.

The event showed the collaboration of ISDN service, equipment, and application developers across the globe. It demonstrated many business solutions for videoconferencing, distance learning, remote LAN access, telemetry, and electronic commerce. These real-life applications highlighted the productivity and cost-effectiveness of using ISDN as the enabling network service.

Because of the success of the Global '95 event, IGICOM plans to sponsor more events. The events, tentatively scheduled for November 1996, will include a series of demonstrations and work sessions, which will become the Global '96 program. Having a series of events allows the activities to become more spread out and focus on specific applications or demonstrations in various world regions. It also reduces the overhead and cost of coordinating a single, all-encompassing event for the year and allows more solution providers to participate.

IGICOM is also planning to fund the development of a document (the *ISDN Implementation Guide*) that will include specific service and pricing information for carriers, along with application notes from each country. This may be aligned with the Global '96 activities. Call IGICOM at the number below for more information about the events and the guide.

There is a great write-up of Global '95 provided by IGICOM that gives the complete details of the event, sponsors, applications, and participants. Read this paperback if you are interested in knowing more about the hard work and business solutions that are growing around the world for ISDN. *ISDN Guide - Global '95* is available from IGICOM at (617)232-3111 (Molly Schaefer) or e-mail at igiboston@aol.com.

Highlights by Country

Those who have observed ISDN use here and in other countries characterize the development of the European market for ISDN as being most developed in the UK, in Germany, and, to a lesser extent, in France. PTT Telecom (Netherlands), Deutsche Telekom (Germany), Swiss PTT, and France Telecom are the main European carriers who have sponsored Euro-ISDN and have sizable ISDN deployment and existing support.

Although these European countries are the strongest supporters of ISDN, it is also spreading to other parts of the world. Let's look at how ISDN is coming along in some specific areas.

ISDN in Canada

ISDN service got a slow start in Canada. During the late 1980s and early 1990s, while the phone companies in the United States were conducting product trials of ISDN, there was little ISDN deployment or testing in Canada.

Since 1992, there has been increased interest in ISDN service, as well as in the data applications that ISDN supports, such as the Internet, online services, desktop videoconferencing, and remote access. Now many carriers in Canada offer ISDN service. Along with Bell Canada, the largest local carrier in the country, Island Telephone in Prince Edward Island, Maritime Telephone in Nova Scotia, BC Telephone in British Columbia, and others provide ISDN service in their respective areas. By 1997, Canada's ISDN service will be close to the service available in the United States.

ISDN in the UK

The UK has an estimated 350,000 to 400,000 ISDN lines installed. The carriers in the UK are primarily BT and Mercury Communications Limited, who have offered ISDN service for the past few years. Euro-ISDN was launched in 1995 for Primary Rate ISDN, and there has been national coverage since early 1995.

The BT implementation has some variation in the signaling (Q931), and some terminal adapter vendors have needed to support a software variant for this slightly different version. BT has announced compliance with the Euro-ISDN standard and continues to improve the availability of Primary Rate ISDN on a limited basis. It is expected that within the next year, there will be one standard version for Basic Rate ISDN and adherence to the Primary Rate standard by both BT and Mercury Communications Limited.

In the past, the demand for ISDN has been voice, and the Primary Rate ISDN demand was for PBX connections. This is not any different than the early demand for Primary Rate ISDN in the United States. The early users of Primary Rate ISDN were large customers who wanted to connect their PBX sites to each other and to the long-distance carrier for long-distance traffic. Now the demand for data applications is growing at a phenomenal rate. The key emerging data application areas are Internet access and remote LAN access.

ISDN in Sweden

Telia AB is the Sweden PTT, and the current number of subscriber lines is just above 6 million. Sweden has a population of around 8.6 million people. The Basic Rate ISDN service is growing about 1000 lines a month. The number of Primary Rate ISDN lines is growing at three to four times that rate, but the base of existing Primary Rate lines is small.

Sweden used a Telia proprietary version of ISDN for many years, but has since converted to the Euro-ISDN standard. This will lead to interoperability and compatibility with other European countries.

ISDN in Switzerland

Swiss Telecom is the PTT carrier in Switzerland. This country has about 4.25 million total subscriber lines. There are nearly 70, 000 Basic Rate ISDN lines and about 4000 Primary Rate ISDN lines.

Like Sweden, Switzerland also had a proprietary version of ISDN since it was introduced in 1989. In 1992, the country started its conversion to the Euro-ISDN standard and full interoperability with the rest of Europe. This is now complete, and Switzerland subscribers can enjoy standard connections with other European countries.

ISDN in Germany

Deutsche Telekom is the PTT carrier in Germany. This company has been sponsoring two versions of ISDN for the past few years: Euro-ISDN and National 1TR6. It plans to convert all ISDN services to Euro-ISDN by the year 2001. Siemens is the vendor of choice in Germany, and its ISDN rates are very close to regular business line service.

As mentioned earlier, Germany has the largest base of ISDN service and will probably keep that distinction for the foreseeable future. Germany also has a very large business community that has ties into the United States—one reason that ISDN is finally catching on in the United States.

ISDN in France

France has around 100,000 to 125,000 ISDN lines installed, and nearly 800,000 lines are forecast by the year 2000. The French companies' investment in ISDN was very aggressive in the late 1980s and early 1990s.

Currently, ISDN service is available in most metropolitan areas in France, and it is expanding into rural areas. As in other countries, the demand for ISDN will increase with the growing popularity of data applications, such as online services and remote LAN access.

NOTE   According to a recent update to the France Telecom home page (www.francetelecom.com), ISDN service (Branded Numberis) is now available nationwide. At the time of this writing, the home page quotes a basic rate for ISDN of about $40 per month plus usage charges.

ISDN in Central and South America

ISDN is being deployed in most Central and South American countries. Over the next five years, it is expected that ISDN will be playing a bigger role, because the economic developments in these areas will require more international connectivity and an overall upgrade to the national network infrastructure.

The investments in Argentina, Brazil, Chile, Colombia, Curacao, and Mexico make up over 50 percent of the telecommunications growth in Central and South America. Argentina, Brazil, Chile, and Mexico represent the majority of this growth. As more and more United States companies increase their business in this part of the world, there will be more investment in digital technology overall, including ISDN technology.

ISDN in Africa

The limited digitalization of public networks in Africa poses a challenge to the deployment and delivery of ISDN to most of the continent. In countries with limited data networks, ISDN provides a key solution to providing a vast and rapid upgrade. By using overlay networks, a standards-based ISDN offering can be made available to areas with the highest demand and need first, in a cost-effective manner. Eventually, the phone companies will upgrade their equipment and integrate the ISDN services into the phone network.

In order to coordinate a standard offering in the continent, a meeting was held in mid-1995 in Namibia, where the adoption of Common Channel Signaling (CCS) using ISUP-based on ITU-TSS was sealed. South Africa began implementing ISDN in April 1995, and this service is experiencing rapid growth throughout the country. South Africa has adopted the Euro-ISDN standard, and it is in the process of formalizing an industry group called the South African ISDN Forum (SAIF) to address ISDN user needs and development issues. The SAIF is working with the IMIMG, and South Africa has a representative on the IMIMG board.

ISDN in Asia and the South Pacific

ISDN is widely available in Asia and throughout the South Pacific. Japan implemented ISDN as early as 1986, and it has been widely adopted for voice applications. ISDN is not widely used for data applications in Asia because PC adoption rates are not nearly as high as they are in North America and Europe.

The most popular applications for ISDN in Japan are videoconferencing, pay telephones, and karaoke machines. The equipment manufacturers from Japan have developed ISDN equipment that supports the videoconferencing and voice features of ISDN and are just now starting to introduce equipment that is more focused on data applications. Much of this equipment will include integrated ISDN terminal adapters or inverse multiplexers that support ISDN. This will dramatically reduce the cost of conversion to ISDN in the near future.

Korea has set up a separate ISDN overlay network, which has created the image of ISDN as a private switched network that requires special equipment to access. The Korean architecture is common to the early ISDN networks in countries that had very little infrastructure in place for digital switched telecommunications.

China and India have just begun to standardize on ISDN and digital technology. This may take some time because of their lack of basic telephone infrastructure. However, during the last three years, there has been a flurry of interest and investment in these two countries (and the countries between them).

Malaysia is another country that is getting some attention from ISDN promoters. New investors and businesses are helping to develop the networks in these countries.

It is important to note that ISDN (and digital telephony) is just one of the telecommunications options being developed in these countries.

ISDN in Australia

During the past five years in Australia, communications companies have been making a large investment in the entire telecommunications infrastructure. At the same time, the businesses of Australia have increased needs for telecommunications on a local, regional, national, and international basis. Bellcore, major telecommunications network equipment vendors, internetworking vendors, and PC manufacturers have seen double-digit growth in volume and sales in Australia. Paul Hogan may have created the tourist boom for Australia, but it was increased international business, including those in high-technology fields, that created the boom in new telecommunications services.

For the most part, ISDN deployment has been concentrated in the major cities, like Melbourne and Sydney. Given that the country is so large, it is not likely that ISDN service will be available outside the densely populated areas of the major cities. But these areas will have 100 percent coverage in their business districts.

ISDN in the Middle East

ISDN is currently in use in Israel, Saudi Arabia, and the United Arab Emirates. As they continue to rebuild cities (and, in some cases, build new cities) that do not have any telecommunications infra-structure, ISDN and digital switches will be the core infrastructure for new digital telephony.

Some of the largest sites in the world that will have digital technology and ISDN as a standard will be in this area. The switch technology and the standard used will most likely be the Euro-ISDN standard, which will increase interoperability with ISDN services in Europe and the United States.

The Worldwide Growth of Data Applications

As we've said, the early ISDN applications in the European markets have been oriented toward voice capabilities and the set of call features that ISDN provides. Now we are seeing a strong trend toward data applications to go along with the voice ones. Videoconferencing and dataconferencing are popular uses, with obvious travel savings for those users.

Internet access, with Web surfing, is a major motive for improving online access. As PC sales increase throughout the world, and many businesses and consumers use the Internet for information sharing, the demand for ISDN data-application support will continue to increase.

The need for remote access to corporate resources and business partners will also grow as LAN connectivity increases. More corporate business end users will demand this service from the increasing number of European and multinational companies.

Many of the internetworking and network-access equipment vendors are looking to the international markets for double-digit growth and increase in the overall market share. Most of these companies already have some international presence but are building the distribution networks and staff to reach more of these countries with as many channels as possible to cover each market or country. This will increase the interoperability between ISDN users in the United States with those in other countries.

The Key Lessons from Foreign Markets

In making ISDN connections in the United States, we can learn quite a few key lessons from the global ISDN experience:

- **Government can influence and foster new technology.** Governments in many of the countries mentioned in this chapter have driven the technology to its acceptance and growth levels by regulations and funding. Although the United States government may not fund the technology and create the standards for service as in other countries, it can provide the support and sponsorship in regulating the service to encourage new digital technology like ISDN. It can be as simple as the current

EUCL (End User Common Line) charge determination and the review of new technologies in different models than older analog technology paradigms.

> **NOTE**
>
> The EUCL charge is a cost to the phone service end user based on the pair of copper wires connected to each user. This charge is a FCC (Federal Communications Commission) allocation method for having the user pay for the ability to access long-distance networks and providers.

- **Standardization increases acceptance.** The European countries had problems until they agreed upon a standard version of ISDN. This allowed interoperability and increased the communication between the countries. Switch vendor manufacturers, carriers (PTTs), and the terminal adapter manufacturers began creating standard functionality. This led to the lower cost of manufacturing and design of the necessary switches and terminal equipment for the ISDN service. The end users benefited from lower cost of equipment for ISDN and the easier implementation of the standardized service.

- **Standardization can limit features.** The Euro-ISDN standard has a limited set of features to minimize configuration issues and allow for easier installation and implementation. This limitation is outweighed by the benefit of having many vendors able to implement one standard and reduce the service cost for all parties.

- **Vendor cooperation leads to standard implementation.** The various vendors of the equipment and the switches needed to agree to standards of service to provide the standard equipment and service elements. The switch providers developed the ISDN service in the standard version to enforce the standard ISDN service with standard feature options for all users.

- **One hundred percent service availability increases public acceptance of any new service.** In the major cities and other populated portions of the countries, end users can have the service no matter where they live. Businesses know that ISDN will be available to all their sites, and most MIS staff do not need to worry about who can get it.

- **Voice applications have some market attraction.** Some predominant voice applications have driven the use of ISDN in the European market. This is not so prevalent in the United States because of the lack of the PBX market coverage. It's more common in other countries because of the lack of network-based Centrex service in the European market. If you never had a PBX or Centrex phone, then the ISDN voice feature set is very attractive. The increased reliability of digital service also has been an attraction to ISDN voice services. In some countries, where PC ownership is not common, it will take time for the data applications to reach the demand we currently have in the United States.

- **Data applications drive ISDN growth.** Data applications collectively, and not one "killer app", are going to drive ISDN growth to higher levels worldwide in the next five years. There is a global trend of all businesses toward more interconnectivity.

In summary, the United States market will see the results of current final standardization efforts for the overseas switch vendors and terminal equipment suppliers. This will increase the acceptance of ISDN for the applications that are the most popular here. If multiple carriers can agree on National ISDN sponsorship, standardized SPID formats, and common service order activation procedures throughout the 50 states, we will have learned our lessons from the international ISDN implementations.

Making Overseas Connections with ISDN

The global use of ISDN will continue as more people do more business together and also as businesses expand into other countries. End users see the need to communicate more often, and ISDN is a network option that facilitates this communication and information sharing.

In the United States, the long-distance carriers have been building more switch infrastructure that supports the needed network connections, with 64-kbps capability. By end of 1995, most carriers will have the network capability for national and international connections to all major business centers in the world. In addition, all major local and long-distance carriers will have some type of additional support structure and centers to assist customers in ordering and implementing ISDN. Many already have established support for international ISDN connectivity. Many will be very knowledgeable about the service, and more important, about the common applications and equipment that are being deployed in various countries.

The United States, through international carriers like MCI and AT&T, and through alliances with the RBOCs and Bellcore, will continue the development of common capabilities so that users will be assured of standards-based international applications critical to enterprise development (such as videoconferencing and remote-data-access features).

All major carriers have made announcements regarding ISDN service availability and support for international markets in the last few months. This is clearly a signal that ISDN support and availability will increase in the next few years. As the globalization of many markets and corporations occurs, the demand for ISDN services—for remote connectivity, desktop conferencing, and online service access—will continue to grow.

The European Conference of Postal and Telecommunications (CEPT) Administrations has developed a *User Handbook for ISDN* that covers Europe and some other countries. The handbook lists the telecommunications companies, contact names, ISDN service overview, and the availability by country. It's a great reference document for anyone who is interested in implementing an ISDN application with business partners or others in Europe.

Because the ISDN implementation and network infrastructure in each country might be different, you need to know which ISDN standard the country you will be communicating with supports for ISDN. Most likely, it will be the National ISDN (the United States standard) or Euro-ISDN. The Euro-ISDN standard has a limited set of features and assumes one terminal per ISDN line. This differs from the National ISDN standard, which can support multiple terminals (eight) on one ISDN line and provides more voice features on each line and each terminal.

The main differences between Euro-ISDN and National ISDN are as follows:

- Euro-ISDN does not use SPIDs, and Euro-ISDN CPE does not undergo layer 3 terminal initialization (see Chapter 3).

- European telecommunications support the concept of "overlap receiving," where one digit at a time is sent from the switch to the customer premises equipment for call termination. (Euro-ISDN also supports en-bloc receiving, where all the digits are sent at once from the switch to the CPE, which is how telecommunications operate in the United States.)

- There are various differences in how protocol errors are handled.

- Euro-ISDN defines additional messages, information elements, and cause values.

- In Euro-ISDN, the SETUP message may contain both keypad and called party number information elements—a definite "no-no" in the United States.

In most cases, you will be able to have at least one B-channel connection to another country. The long-distance carrier and the terminal adapter will determine the interconnection limitations of your ISDN service.

PART II

What Can I Do
with ISDN?

5

About Online Information Access

- Online resources for "information at your fingertips"

- Advantages of ISDN connections to the Internet and to online services

- Internet service providers and ISDN services

- ISDN hardware for online information access

- Online information access applications for ISDN

- Some case studies of online information access with ISDN

- Online information tariff considerations

These days, you'll find 28,800-bps modems everywhere. Those who have been online for awhile are noticing something different for the first time. In the past, a new modem was always just on the horizon, offering an increase in speed ranging from 50 to over 100 percent. If you were an "old-timer" using a 300-baud modem, you could always see a faster modem advertised for a future release—1200, then 9600, then 14,400 baud. Today, owners of 28,800-bps modems have only the 33,600 bps to look forward to. What happened? It's the end of the line. The theoretical maximum for analog modems is around 30,000 bps, without compression. Where do you go to get more speed? ISDN is here and ready to go today.

In this chapter, we will examine several online services, the ISDN design issues related to online information access, online applications that can benefit from ISDN, some case studies, and some tariff considerations.

An Overview of What You Can Find Online

Everyone is talking about the "Information Superhighway." Why? What exactly is it? New services, applications, and technologies are multiplying, almost as fast as the number of users deciding to join the "network of networks." Before we talk about the benefits of traveling on the Information Superhighway's version of the express lane—ISDN—let's go over some of the basics of being online and what it means to you.

What Is the Internet?

A computer network is simply two or more computers connected by wires. Computer networks allow interconnected users to share printers and files. When one network is connected with another, an internet (lowercase *i*) is formed. The Internet (uppercase *I*) is the international network of interconnected computer networks.

Estimates of the number of individuals on the Internet vary widely, but approximately 50 million use the Internet worldwide, making the Internet the world's second-largest communications network, after the POTS (the plain old telephone) network. One notable difference between the Internet and the telephone network is that electronic mail (e-mail) sent to users outside your home country typically costs the same (at least for the end user) as e-mail sent to users within your home country. As a result, individuals from all over the world can meet on the Internet in virtual communities.

Buzzwords like the Information Superhighway, cyberspace, and the national information infrastructure are nicknames for the Internet, but they are not helpful to understanding the Internet. The Internet is the catch-all word used to describe the massive worldwide network of computers. The word "internet" literally means "network of networks." In itself, the Internet is comprised of thousands of smaller regional networks scattered throughout the globe. On any given day, it connects roughly more than 20 million users in more than 50 countries.

Nobody "owns" the Internet. Although there are companies that help manage different parts of the networks that tie everything together, there is no single governing body that controls what happens on the Internet. The networks within different countries are funded and managed locally according to local policies.

The phrase "having access to the Internet" usually means that one has access to a number of basic services, such as e-mail, interactive conferences, information resources, network news, and the ability to transfer files.

What Is the World Wide Web?

For 50 years, people have dreamt of the concept of a universal database of knowledge—information that would be accessible to people around the world and provide links to other pieces of information. It was in the 1960s when this idea was explored further, giving rise to visions of a "docuverse" that people could swim through, revolutionizing all aspects of human–information interaction. Now technology is catching up with these dreams, making it possible to implement concepts, introduce products, and enter markets on a global scale.

The World Wide Web (also referred to as the Web, WWW, or W3) is officially described as a "wide-area hypermedia information retrieval initiative aiming to give universal access to a large universe of documents." What the World Wide Web project has spawned is a tool to provide users on computer networks a consistent means to access a variety of media (text, audio, video, and animation) in a simplified fashion (point and click). Using a popular software interface to the Web called a *browser*, the Web project has grown into a vehicle that is changing the way people view and create information; it's the first truly global hypermedia network.

The World Wide Web is often referred to as the Internet, but the two are not the same thing. The Web refers to a body of information, an abstract space of knowledge; the Internet refers to the physical side of the global network, a giant mass of cables and computers.

The World Wide Web uses the Internet to transmit hypermedia documents between computer users internationally. Organizations or individuals are responsible for the documents they author and make available publicly on the Web. Via the Internet and the World Wide

Web, hundreds of thousands of people all around the world have information available from their homes, schools, and work places.

NOTE The Web began in March 1989, when Tim Berners-Lee of CERN (a collective of European high-energy physics researchers) proposed the project to be used as a means of transporting research and ideas effectively throughout the organization. Effective communications were a CERN goal for many years, since its members were located in a number of countries. Now the Web helps "members" of the worldwide human race to communicate.

Nobody knows exactly why the World Wide Web is so popular. The Web is growing so quickly that as soon as experts come close to getting a fix on it, it grows out of their "sites." Currently, we can only make intelligent guesses or estimates to describe the growth of the Web in terms of users and Web sites.

The popularity of many graphical user interfaces (GUIs) for browsing ("surfing") the Web has made the hard-to-master resources on the Internet, well, user friendly. It is probably this ease of use and the open access that has caused the explosion of Web traffic in 1993. This explosive growth continues today with some estimates maintaining that the World Wide Web doubles in size every *two months*! How long will this continue? When or will it end? Again, nobody knows.

Five Internet Tools

The Internet has its own toolkit for digging or surfing for information. Here, we'll describe five main tools: e-mail, FTP, Gopher, World Wide Web, and WAIS.

Electronic-Mail (E-Mail)

E-mail is a tool that allows one user on the Internet (or other online services) to send a message to another user on the Internet. An e-mail message may contain text or pictures and sound encoded as text, but most often it is plain text. E-mail programs are the most widely used Internet tools, because the Internet is primarily a communication medium between users.

Not all e-mail users need to be human. Some e-mail is generated from automated programs that can send e-mail messages to a group of individuals interested in the same type of information. This is similar to an electronic newsletter, which allows all of the subscribers to submit articles or comments for immediate publication. By redistributing your e-mail message in this way, the automated e-mail program creates a virtual community, often referred to as a discussion group. This automated program or system is known as a Listserv.

The Listserv family of automated programs allows individuals to subscribe to various lists (or discussion groups). The Listserv program handles all the administrative tasks (such as adding or deleting individuals from the subscription list and redistributing e-mail to all of the list's subscribers), leaving individual subscribers free to discuss substantive issues.

File Transfer Protocol (FTP)

FTP is a tool that allows users on one computer (the local computer) to connect to another computer (the remote computer) for the limited purpose of copying files from (and/or to) the remote computer. A computer set up to accept incoming FTP requests from another computer is called an *FTP server*.

Usually, the administrators of an FTP server will copy certain files to a public directory on the FTP server. In this way, information is made available to the Internet community. An FTP server is like a bulletin board. The owner of the FTP server can add and delete files from the public directory on the server just as notices can be physically tacked to or removed from a bulletin board.

Gopher

Gopher is named for the mascot of the University of Minnesota, where it was developed. Gopher is a menu-driven program, much like an ATM at a bank. The Gopher server, a computer set up to run the program, is configured with a main menu and a series of submenus. By selecting a particular menu item, you can view documents, run other Internet programs, or connect to another Gopher server.

Because one Gopher server can connect to another, users can look at menus and submenus from Gopher servers all over the world. When you connect to another Gopher server, the Gopher program on your local computer copies the remote menu. This feature allows many Internet users to view the same Gopher menus. This is analogous to signing a book out of the library one page at a time instead of making others wait for the whole book. While many organizations are still publishing information via Gopher servers, World Wide Web servers are today's primary publishing platform.

World Wide Web

A Web server is a computer set up to run the World Wide Web program. Unlike Gopher, which presents you with a series of menu items, the World Wide Web presents the user with documents. Like a Gopher menu, each World Wide Web document can contain links, which often appear as underlined text or images. When you select a particular link, you can view documents, run other Internet programs, or connect to another Web server. The home page for a Web server is analogous to the main menu for a Gopher server.

To access the Web, you run a browser program that can read and retrieve documents. Mosaic, Internet Explorer, and Netscape are the most popular Web browser programs. The browsers can access information via FTP, Telnet, Usenet, Gopher, Wide Area Information Servers, and other sources.

Wide Area Information Servers (WAIS)

WAIS is a networked, full-text, information-retrieval system developed by Thinking Machines, Apple Computer, and Dow Jones. WAIS currently uses TCP/IP (the Internet networking protocol) to connect client applications to information servers. Client applications are able to retrieve text or multimedia documents stored on the servers.

With WAIS, client applications request documents using keywords. Servers search a full-text index for the documents and return a list of documents containing the keyword. The client may then request the server to send a copy of any of the documents found. The WAIS software distribution is available via anonymous FTP (a service that allows any Internet user to download files from a computer connected to the Internet, without first establishing an account to use that computer). The World Wide Web is also swallowing up WAIS. Web search engines are becoming the dominant use of WAIS.

Who Needs Access to Online Information Services?

Who needs to get online? The answer is just about everyone. In today's world, everyone must find a way to keep current with new developments, a changing work place, and a globally influenced marketplace. Let's look at a few features available to today's online community.

Are you interested in art and architecture? On the Web, you can find art galleries and famous people, such as Frank Lloyd Wright and Leonardo da Vinci. Tour online museums, such as the Louvre in France.

The intellectually inclined can visit sites on literature, philosophy, and science. The Web offers links to libraries, including the Library of

Congress. College students should prepare—doing research on the Web cuts down on trips to the library. Gee, what will you do with all the extra free time?

Do you like to play games? The Web has games you can play alone or against other netizens. Hate to lose? Yes, they even have Web sites with secret tips and hints. We know, some people call it cheating (maybe that is why your friends always win!). Or instead of playing yourself, you can look up your favorite sport or team.

Maybe you want to rent a movie. Try the Internet Movie Database or the MovieWEB, which previews coming attractions before they are even released. Are you interested in films? Maybe the Horror Web page is for you. If you are a little old-fashioned, check out the Silent Movie Web page. Or perhaps you would just like to check out the movie you saw advertised in the paper. Now most movies have their own Web page. But wait! You may want to get ISDN first, because the movie clips can be large. How does 5 megabytes sound? Hmm, that is about 23 minutes using a 28.8-kbps modem, but less than 7 minutes with ISDN.

But going online is not just for having fun. Let's look at some more serious applications.

The Web offers pages to track your portfolio or research companies. More and more are offering "personal" information services, tailored to your interests. Will new laws or regulations affect your investments? You will find a number of sites dedicated to politics and government. You can even visit the White House.

Get the facts before we elect our next president. For example, acting under the directive of the leadership of the 104th Congress to make Federal legislative information freely available to the Internet public, a Library of Congress team brought the THOMAS World-Wide-Web-based system online in January 1995. The first database made available was Bill Text, followed shortly by Congressional Record Text, Bill Summary & Status, Hot Bills, the Congressional Record Index, and the Constitution. Enhancements in data and search and display capabilities have been continuously added for each database.

A number of banks are providing services online, so you have someplace to put all the money you are making with your investments.

Does all this talk of money and politics make you feel ill? Or are you concerned about your physical, as well as, financial health? The Web offers links to health and happiness (a Web page on Prozac?). The number of opportunities are growing every day—so many, you could write a book (and some people have). But remember, movie clips, music samples, games, and exploring virtual worlds all need more bandwidth. This bandwidth can be provided by ISDN, which is what this book is about. The next section will get you up to speed on the terminology associated with online services, and then we'll return to some specific ISDN issues.

Online Buzzwords Brief

Here are some terms you may come across in discussions of online information access:

- Your *account name* is the eight-character name you choose when you sign up with an ISP.

- *Anonymous FTP* is a widely implemented service that allows any Internet user to download files from a computer connected to the Internet, without first establishing an account to use that computer.

- *Archie* is a service that tells Internet users where to find files by maintaining a database of what is stored at anonymous FTP sites around the world.

- An *archive file* (when speaking of a file from the Internet) is a file that has many files within it. This process is similar to placing many small items in a bag, so that instead of moving many little items, you move one large one. The archive file usually ends with the extension .ZIP or .ARJ (this extension indicates which program was used to create the archive). This is important because in order to "unpack" these files, you must have the accompanying program (for example, PKUNZIP for .ZIP files). A self-extracting archive is one that ends in the extension .EXE. To use these files, you only need to run the .EXE file, and the archive will unpack itself.

- *Attachments* are usually associated with e-mail. You are able to send, or "attach," text and/or binary files to e-mail messages. The specific method for attaching files differs from one mailing program to another. The actual step-by-step instructions should be accessible from the Help menu of your mail software.

- A *BBS* (bulletin board system) is a dial-in or Telnet-in computer that usually provides e-mail, file archives, live chats, and other services and activities. Many are connected to the Internet.

- *Bookmarks* allow you to save a reference to a place on the Internet so you can find it again. Originally used with Gopher, it is now a popular feature of most Web browsers. Some FTP programs also allow you to save bookmarks (such as Anarchie for the Macintosh).

- A *client* is the machine that is placing a demand on another machine. Your computer is usually the client. For example, if you are using the Netscape browser, when you "go to" another location, you are the client placing the demand to see a Web document on the host.

- *Cyberspace* is the electronic "world" of computers and networks. Author William Gibson coined the term in his novel *Neuromancer*.

- A *domain name* is how you describe the name that is to the right of the @ symbol in an Internet address. For example, Sybex.com is the domain name for Sybex.

- *Domain name system* (DNS) is the naming system that identifies computers connected to the Internet.

- The term *download* refers to the process of getting information from a remote computer to your local computer. Since you dial UP to a computer, you "DOWNload" a file from that computer.

- Your *e-mail address* is your account name, followed by the @ symbol, followed by your domain name. For example, if you are a commercial user and the domain name is "star" and your account name is "jsmith," your e-mail address would be jsmith@star.com. If you are part of an educational institution, your address would end with ".edu," as in jsmith@star.edu.

- *Eudora* is an e-mail program based on POP mail. Eudora is a commercial program that can be loaded on a desktop computer. It provides a graphical interface for users and can download messages from and upload messages to an Internet host supporting POP mail.

- *FAQ* (Frequently Asked Questions) is a list of questions and answers related to a newsgroup, software, Web sites, and so on. FAQ lists prevent newsgroup discussions from being overrun by common user questions.

- A *firewall* is a computer or another piece of hardware or software set up to manage traffic between an Internet site and the Internet. It's designed to keep unauthorized users from tampering with a computer system or network.

- A *flame* is basically some form of derogatory communication directed at another person, and a *flamer* is the author of such a comment.

TIP

Newbies (newcomers to the Internet site) are favorite targets of flamers, just because they are newbies. Flamers will criticize and belittle anyone and everyone just to hear themselves speak. People who flame others are totally inconsiderate of the other readers and should be considered the lowest of the low life. "FLAMERS ARE NOT WORTHY TO BE ON THE INTERNET." As you can see, the use of all capitals is considered yelling and in bad taste for anything other than a few words for emphasis. If you are a newbie, read the FAQ list before asking questions in any newsgroup. It will often answer your particular questions, and you will also avoid being flamed.

- *FTP* (File Transfer Protocol) is a common protocol used for downloading files via the Internet.

- *Gopher* is a menu-based information system running on the Internet. Gopher is a client/server application developed at the University of Minnesota and distributed freely across the Internet. Many Internet users have taken advantage of this system to make available a wide range of information and provide access to diverse computing resources in the form of databases, software, and so on.

- *Gopherspace* refers to all the Gopher servers on the Internet, because all the servers can connect to each other, creating a unified "space" of Gopher menus.

- A *host* is a computer connected directly (essentially permanently) to the Internet. This term is typically used to refer to machines that provide account, database, or archival services.

- *HTML* (Hypertext Markup Language) is the standard coding language in which World Wide Web documents (Web pages) are written. HTML has two versions defined with a third being finalized (called HTML3). Some older Web browsers may not be able to view Web pages written using the newer versions of HTML. To further enhance HTML, Netscape has also added its own codes, which may or may not ever be integrated into the standard HTML. Microsoft has also come up with its own codes, which will only be readable using Microsoft's browser (Internet Explorer).

- *HTTP* (Hypertext Transfer Protocol) is the protocol that is used to send and receive Web pages over the Internet.

- *Hypertext* refers to documents that provide links to other related information. The Web is based on hypertext documents.

- An *Internet number* is the unique numeric address, or *dotted quad*, of the form 123.456.7.89, that identifies every system attached to the Internet.

- An *IP address* (or just *address*) is a unique address assigned to each computer on the Internet. This address is used to deliver e-mail, connect to remote computers (Telnet), and transfer files (via FTP).

- The *IRC* (Internet Relay Chat) is the live-chat area of the Internet in which real-time conversations among two or more people take place in virtual rooms. In some virtual rooms, you have to be a fast typist and a speed reader to get a word in edgewise.

- An *ISP* (Internet service provider) is an organization that, for a fee, provides a connection to the Internet.

- *Java*, formerly known as Oak, is an object-oriented programming language. Using Java, developers can write custom miniapplications called Java applets, which will provide Internet sites with a huge range of new functionality: animation, live updating, two-way interaction, and more. Java applets allow cross-platform programmability and can be embedded right into HTML pages.

- *Listserv* is a software program for setting up and maintaining discussion groups. Many Listserv discussion groups are a gateway to Usenet newsgroups.

- *Local computer* or *local system* refers to the machine that accesses the Internet. Usually, your home or office computer—the machine in front of you—is the local computer.

- *Lycos* is a database of more than 3.5 million Web sites. This widely used search engine allows you to set custom search configurations to help you find what you are looking for. Lycos is found at http://lycos.cs.cmu.edu. Another favorite search engine is Yahoo.

- *MIME* (Multipurpose Internet Mail Extension) is an extension that lets you transmit nontext data (such as graphics, audio, and video) via e-mail. Many e-mail packages such as Eudora now automatically use the extension when transmitting any e-mail.

- *MUD* (Multi-User Dungeon) are role-playing games that were originally modeled on Dungeons and Dragons games, but are now used as conferencing tools and educational aids and in other contexts.

- *Netiquette* refers to the unwritten but widely accepted rules of proper behavior on the Internet.

- *Newbie* is someone who has little or no experience using the Internet. A person can avoid the tag newbie (and often an accompanying flame) by staying quiet and reading every FAQ they can before asking anything.

- A *newsgroup* is a group devoted to a particular topic of Internet news, a widely used Internet service. You can find newsgroups that concentrate on academic subjects, current events, business, hobbies, and many other special interests. There are newsgroups devoted to more than 10,000 separate topics.

- *POP* is an acronym for Post Office Protocol. E-mail programs that use the POP system run on a desktop computer and retrieve messages from the computer where they are stored upon receipt. E-mail programs designed in this way can provide users with a graphical interface similar to other programs already in use on their desktop machines. Eudora is an example of a popular POP-based mail program.

- *PPP* (Point to Point Protocol) is the communications protocol used over serial lines to support Internet connectivity. It is used in conjunction with TCP/IP and is similar to SLIP but usually more reliable and faster.

- *Remote computer* or *remote system* usually refers to the machine you connect with from your local computer. In the case of Internet service providers, the machine on the other end of your modem is the remote computer. If you are using FTP, the machine that you connect with is the remote computer.

- A *search engine* is a tool for finding information on the World Wide Web. In most cases, you enter the words you want to search for, and the search engine looks for those words in a database. It then displays a list of URLs that match your request, with links to each one. Yahoo and Lycos are examples of search engines.

- *SLIP* (Serial Line Internet Protocol) is a communications protocol used over serial lines to support Internet connectivity. It is used in conjunction with TCP/IP.

- A *SYSOP* (systems operator) is the person who does day-to-day maintenance of a BBS.

- *TCP/IP* (Transmission Control Protocol/Internet Protocol) is a set of protocols that allows one computer to talk to another and one network to talk to another.

- *Telnet* is an Internet application that lets you log in to computers around the world that are connected to the Internet, and to use them as if they were your own. Usually, knowledge of Unix is required to utilize Telnet fully.

- *TLAs* (three-letter acronyms) are the mysterious terms used to refer to things net-related. Examples are FTP, URL, and WWW. Of course, many acronymns now have four or more letters, but they are often still referred to as TLAs.

- *Unix* is a type of operating system that is commonly used on computers connected to the Internet. Many Internet users access that network through accounts on computers running this operating system.

- *Upload* refers to the process of getting information from a local computer to your remote computer. Since you dial UP to a computer, you "UPload" a file from your local computer.

- A *URL* (Uniform Resource Locator) describes the access method and location of a resource on the Internet. Most Internet sites have URLs. URLs have the format of *access method://location*. For example, to access the homc page of Sybex, your Web browser uses the URL http://www.sybex.com. With many Web browsers, the URL will appear underlined and have a different color than the text surrounding it. Other access methods include FTP, Gopher, and Telnet. (You can even use the access method of *file://* to open files stored on your own computer.)

- *Usenet* is a collection of newsgroups organized by topics. This term is also used to describe a network of computers, not all of which are on the Internet, that receive Usenet newsgroups.

- *Veronica* is a database for searching for information within Gopherspace using keywords and subjects. You can access Veronica at gopher://veronica.scs.unr.edu.

- *Virtual Reality Modeling Language* (VRML), pronounced "Ver-Mil," is a text-based, 3-D description language. Like HTML files, VRML documents are ASCII text files that describe what you want to display in the browser. Unlike HTML, which is a text "mark-up" language, VRML commands describe three-dimensional objects, lighting, movement, and similar items.

- *WAIS* (Wide Area Information Server) is a tool to search through indexed databases based on keywords and retrieve any related documents found.

- A *Web browser* is a program that allows you to access the Web and view the Web pages. The most common Web browsers are Netscape, Internet Explorer, and Mosaic.

- *Yahoo* is a Web search engine that lets you search by keywords or by using a tree-based menu system. Yahoo is at http://www.yahoo.com. Another favorite search engine is Lycos.

Advantages of Using ISDN for Online Services

With the increase of bandwidth-intensive graphical content on the Internet, users can dramatically improve the performance and speed of their Internet connections with ISDN service. In addition to faster World Wide Web access and file transfers through the Internet, you'll have better connections to online services, such as America Online, Prodigy, Microsoft Network, and CompuServe. These online services are either updating their Internet components or are striking deals to replace the old and tired browsing programs.

ISDN's speed, quality, and fast connect times mean the following to those connecting to the Internet and online services:

- Less time waiting for graphics and files to download
- Faster navigation through graphical user interfaces (GUIs) made with colorful graphics and buttons
- Shorter connect times (about 3 to 5 seconds)
- Higher-quality connections with less noise
- Reduced number of retries to connect

Less Time Waiting for Files

ISDN is your ticket to spending less time waiting and more time seeing all of the Web's wonders. Web sites may feature games, background information, prizes and sweepstakes, and movie trailers.

Unfortunately, all this fun involves images, and images gobble up bandwidth.

For example, some movie trailers can be huge, something like 12 megabytes. For 28,800-bps modem users, a 12-megabyte file takes approximately 55 minutes to send, assuming the site is not experiencing too much traffic (lots of users, like you, all trying to request access to the Web server). ISDN access, on the other hand, cuts big files down to size. You can download that 12-megabyte file in less than 14 minutes!

Faster Navigation through GUIs

As more and more people discover the Web, corporations and services are finding themselves competing to win your attention. To raise their site's appeal and effectiveness, Web page designers are focusing on colorful graphics, dynamic applications, and even sound and music to grab and keep your attention.

As we've said, the price is large files and pages filled with graphics that you must wait to see. You will quickly discover that even beautiful, dynamic Web pages become tedious as each page turn means another wait.

Get On and Stay On

ISDN is digital and, like your music CD, provides clean and clear transmissions. Some ISDN equipment can be configured to give you a quick connection, in less than 5 seconds; a modem's handshaking can take 30 seconds or more.

When ISDN equipment is provisioned properly, you can take advantage of one B channel, and have the equipment dynamically add another should your online browsing hit a graphically rich site. We will cover these provisioning features in Chapter 10.

ISPs and PRI

Behind the scenes, the explosion of interest in the Internet has Internet service providers (ISPs) scrambling to install the latest and most comprehensive networks available to meet the demands and needs of their customers. "Information at your fingertips" is a service demanded by businesses and individuals. Students may now complete research projects by surfing the Web. Businesses are conducting online marketing research and launching campaigns.

One industry trend is the use of Primary Rate ISDN (PRI) service to create a backbone network for ISPs. Today, ISPs are able to provide a local phone number for a city or neighborhood, ranging from analog modem rates of 14.4 kbps to digital rates up to 128 kbps. A user only needs to dial the local phone number supplied by the ISP's PRI service, thereby grabbing a B channel or two. The PRI line carries the traffic back to hub sites. Additional traffic is aggregated further and routed to remote Internet servers. The process in analogous to a stream's tributaries flowing into a river that eventually reaches the ocean.

PRI enhances the ISP's service by providing extremely fast call setups and digital-quality transmission. The fast call setups allow the ISPs to pump more data across the network than they can with traditional methods. Also, PRI aggregation allows ISPs to save money by avoiding the rack of modems required for analog service.

Design Considerations for Online Information Access

We will look at the following design issues related to online information access using ISDN services:

- Determining ISDN availability

- Choosing an access provider

- Selecting your ISDN hardware

Is ISDN Available?

Before you can take advantage of ISDN's speed, quality, and quick connections, you need to determine ISDN availability in your area. This sounds like an obvious step, but most people assume that ISDN is available everywhere. Well it is getting close, but it is not there yet. So, before you do anything else, you should contact your local phone company to determine if ISDN is available in your area. See Chapter 9 for more information about checking ISDN availability.

WARNING One of the first traps that many users fall into is ordering ISDN service from their local RBOC, first. Most of us are conditioned from our analog days to ordering phone service and then hooking up a phone. To connect to the Internet using ISDN, you work backward. First, you decide on an application, in this case, an Internet connection. Next, choose an Internet service provider that offers ISDN or supports ISDN connections. Then select your ISDN equipment (a terminal adapter). After you've taken all these steps, you are finally ready to approach your local phone company.

Which ISP Should You Use?

Before you shop for an ISP, you should answer these questions about your online activities:

- How much time do you spend online?

- What type of services do you use?

- What type of applications do you use?

The answers to these questions will help you choose the type of service (fixed versus measured), based on how you plan to use your online service.

If you know what you need, you will be better prepared to evaluate different service plans (discussed later in the chapter, in the section about tariff considerations). The following is a summary of questions to ask a prospective ISP before signing up:

- What bandwidth is supported (64 or 128 kbps)?
- Does the ISP support compression? If so, what kind (Microsoft, Ascend, or Stac)?
- What type of equipment does the ISP support or recommend?
- Is the ISP willing to share references from ISDN customers?

Keep in mind, if you pick a successful ISP, its support team has probably helped hundreds of users get connected. They will understand many of the first-time experiences (or problems) you may encounter.

Another option is to subscribe to a commercial online service, which may be the quickest and easiest way for you to get on the Internet. Currently, the six largest national commercial online services are Prodigy, CompuServe, America Online, Microsoft Network, GEnie, and Delphi.

Internet Speed

The growth in Internet traffic is a concern to many groups. The original design for the Internet was as a widely dispersed communication system that would be difficult to knock out during a nuclear strike. The Internet was never designed or intended for the amount of traffic it handles today.

Experienced Internet surfers know that no matter how fast your connection speed, you are limited by several factors:

- The protocol (handshaking) we currently use
- Traffic trying to access a popular site
- The speed of the communication link to the remote site
- The speed of the server answering your request

ISP technicians, whose position gives them a unique opportunity to connect to the Internet through very high-speed connections, have found that their speed, under optimal conditions, usually tops out around 200 kbps to 300 kbps.

The Internet is like our nation's freeway system. If you're driving a Ferrari, you'll find it difficult to travel on Interstate 5 at speeds that reach the vehicle's maximum.

What Hardware Do You Need?

After you choose your ISP, before you buy any hardware, be sure to ask your ISP what brand(s) of terminal adapters the service supports. You will find this research very helpful when choosing hardware and software. By staying with compatible equipment, you make an ally of your ISP. Later, if you run into problems with your equipment or configuration, you will have a resource to help you sort through them.

NOTE   Most ISDN ISPs are using Ascend's Max 4000 because of the router's ability to support compression and Multilink Point-to-Point (MLPPP) connections and BONDING. Of course, if you plan to use only one B channel, which is what most online services support, you won't need to worry about BONDING and MLPPP schemes that allow you to combine B channels to create more bandwidth.

When you shop for ISDN equipment, besides compatibility with your ISP or online service, you will need to decide if you want an internal or external terminal adapter.

An internal adapter provides a direct connection to the computer's bus. This will help you reach the maximum bandwidth available through ISDN (128 kbps). Also, you will not need to dedicate a serial port on the back of your computer to handle ISDN communications. However, using an internal model has two trade-offs.

- Because ISDN requires an external AC power supply, you will lose your access to your ISDN line when you turn off your computer. You will need to have a backup analog phone service to make phone calls when your computer is off.

- As is the case with any internal board, you may find yourself dealing with the added complication of finding an available interrupt. In some cases, you will need to disable an external serial port to avoid conflicts with your internal terminal adapter.

If you choose an external terminal adapter, you can avoid potential conflicts with an internal board and you will have access to your ISDN phone service, even when the computer is turned off. Unfortunately, the external adapter has a trade-off in speed.

Today's IBM PC compatibles and Macintosh computers do not have a serial port that is able to handle the high speeds of an ISDN connection. If your computer is an older model, you may even need to upgrade your serial port to a high-speed UART 16550. However, even the high-speed serial port is capable of transmission speeds up to only 115 kbps. In addition, the serial port still uses a stop bit, a remnant of asynchronous transmission schemes days. Digital technology such as ISDN makes stop bits superfluous.

This additional overhead usually slows your Internet access connection down to 92 kbps. Some quick math will show that you get to use less than half of your second B channel, or 28 kbps. You will want to keep this in mind while shopping for an ISDN ISP that charges by the B-channel minute (see the section about online information access tariff considerations, later in the chapter).

ISDN Applications for Online Information Access

The trend in online applications is toward high-bandwidth technologies that benefit from the larger "pipe" provided by ISDN. Here, we will cover four applications that represent this trend:

- Java applets
- Shockwave
- Virtual Reality Modeling Language (VRML) applications
- Avatar chat applications

What Are Java Applets?

Java applets are small programs written in the Java programming language. Using Java, Web page designers can create pages that include animation, graphics, games, special effects, and make Web pages that traditionally display information become interactive. With a mouse, keyboard, and user-interface tools (such as buttons, sliding bars, and text fields), users can interact with the content provided by a Java-powered Web page. All this "action" can happen because Java applets are embedded within the Web page. After users access a Web page, the applets are downloaded to their computers and executed.

A Short History of Java

Java began life as a programming language intended for use in consumer products. The Java programming language was designed and implemented by a small team of people at Sun Microsystems in Mountain View, California. It shares many superficial similarities with C, C++, and Objective C (for instance, FOR loops have the same syntax in all four languages), but it is not based on any of those languages, nor have efforts been made to make it compatible with them.

As Java evolved through the application to several projects, the World Wide Web was beginning to move from a text-based interface to a more graphical interface, generating lots of interest. During the Web's development, the Java development team realized that Java would be an ideal programming language for the Internet because of its ability to run on many types of computer programs.

Java Nuts and Bolts

Because Java is a programming language, we felt it would be easier to understand Java by offering an example of what can be done with it. If you follow company stocks or like to be your own broker, you will find several low-cost alternatives to guide your investments. WallStreetWeb by BulletProof is one of the new breed of low-cost Web applets that offer services found in expensive software packages. For $9.95 a month, WallStreetWeb, shown in Figure 5.1, handles stock quotes, charts, searches, and portfolio management. WallStreetWeb's home page is at http://www.bulletproof.com.

WallStreetWeb's interactive stock charting

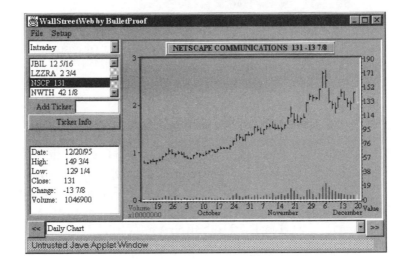

Developers are creating many new Java applications. You may want to follow Java Applet Rating Service (JARS) to keep abreast of new developments. You can find this Web site at http://www .jars.com/. Figure 5.2 shows an example of this Web site.

FIGURE 5.2:

JARS: The Java Applet Rating and Review Service on the Web

What Is Shockwave?

Shockwave is a key component of Macromedia's solution for moving creative professionals who develop digital media to the World Wide Web. It enables them to use their existing authoring tools, skills, and content immediately.

Those who use one of Macromedia's authoring tools (Director, FreeHand, or Authorware 3.5 for Windows or Macintosh) can create a "shocked" piece for the Internet or in-house intranet. Users simply create the piece, compress it using Macromedia's compression software (called the Afterburner) to optimize the piece for Web delivery, and upload it to their Web page. That's it! You've got interactive media on the Web.

Using Shockwave tools, developers can produce Web sites that can be changed easily and rapidly, and that can be customized based on context and events. With Shockwave software, anyone using the Netscape Navigator 2.0 browser can view interactive multimedia, including animation, sound, and high-resolution scalable and zoomable digital art.

A Short History of Shockwave

Macromedia introduced Shockwave in December 1995, when it unveiled Shockwave for Director 4. Overnight, it revolutionized the Internet by bringing interactive multimedia to the World Wide Web. In April 1996, Macromedia introduced new Shockwave software with expanded abilities to display additional forms of interactive multimedia, learning, and digital arts to the World Wide Web.

Since its introduction, more than a million people have downloaded and used Shockwave to view thousands of "shocked" Web sites. Shocked contents fit into the dynamic Web sites created by the Backstage tools that Macromedia recently acquired from iBand.

Shockwave Nuts and Bolts

Because Shockwave software operates on standard Macromedia files, the features of Macromedia's products for creating multimedia and interactive graphics applications are immediately accessible to users of Macromedia authoring tools.

Shockwave software is made up of two distinct parts: the Shockwave plug-ins and the Afterburner applications. The Shockwave plug-ins are a set of viewers for Macromedia datatypes that are packaged as plug-ins to the two major Web browser architectures, and which play back Macromedia titles on the network. The Afterburner application is a Macromedia Xtra or separate utility program that compresses and optimizes standard Macromedia files for use on the Web. Separate plug-ins and Afterburners are currently available for Director, FreeHand, and Authorware, and for different platforms of Netscape Navigator 2.0, including 16-bit and 32-bit on Microsoft Windows, and 68K and PowerPC on the Apple Macintosh. Shockwave plug-ins packaged as Active Objects for use with Microsoft's Internet Explorer will be available soon.

Shockwave solutions will be available for all of Macromedia's authoring products before long. Shockwave will be ubiquitous in that it will be available for common computer platforms and for the most widely used Web browser architectures.

The following are some examples of what developers have produced with Shockwave tools:

- Online advertising (Director)
- Games (Director and Authorware)
- Presentations for communicating marketing and product messages (Director and Authorware)
- Computer-based training and educational courseware including quizzes and tests (Authorware)

- Interactive reference works and publications (Authorware)

- Kiosks for both communication and user input (Director and Authorware)

- Surveys and questionnaires (Authorware)

- Maps and site plans (FreeHand)

- Online technical illustrations and product catalogs (FreeHand)

- Presentation of illustrations, logos, artwork, and other graphic images (FreeHand)

- "Highlight" animations like "flying logos" that add interest to Web pages (Director)

Shockwave is available free of charge on Macromedia's Web site at http://www.macromedia.com/. The Shockwave plug-ins can be downloaded in a single package ("Easy Installation"), or the user can choose which Shockwave product version and platform to download. Each of the Afterburner compression programs is also available for downloading.

What Is VRML?

VRML is software that creates three-dimensional virtual reality worlds for the World Wide Web. VRML is a text-based, three-dimensional description language. VRML browsers or plug-ins let you fly through interconnected, three-dimensional worlds on the Internet.

The VRML browser can manage three-dimensional tasks, such as progressive rendering, physics-based navigation, collision detection, animated viewpoints, and image backgrounds.

A Short History of VRML

In 1994, research into "sensualized" interfaces began to receive attention from the press and the industry. Virtual reality became a term to define a wide range of technologies and began to instrument a fundamental change in the approach to designing user interfaces, moving it from a desktop metaphor to a human-centered design. The space around the user became the computing environment, and all of the user's environment and senses were engaged in the interface.

Because of recent developments in the Web at the time, several efforts began to make the World Wide Web more perceptual. Or as Mark D. Pesce, one of the founders of VRML, states "to put the space in cyberspace."

VRML Nuts and Bolts

Using VRML, you can create your own three-dimensional worlds. You can start by building a virtual world, such as virtual rooms, buildings, cities, mountains, oceans, and planets. The next step is to fill your world with anything that suits your imagination: virtual furniture, cars, people, spacecraft, and so on. Figure 5.3 shows an example of a virtual world with a virtual model of Stonehenge.

FIGURE 5.3:

A scene from traveling around a virtual Stonehenge

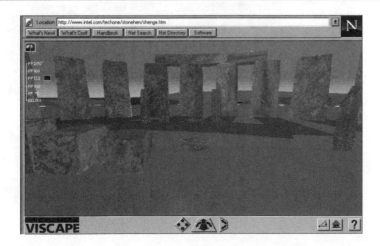

What Are Avatar Chat Applications?

The process of "chatting," or talking to others on the Internet, is as old as the Internet itself. Today, a new approach to sharing and conversing with others, known as *avatar chat,* is available on the Internet. Instead of simply exchanging text with each other, you choose an image that represents you—an *avatar*—such as a horse, an angel, a fish, or whatever you like. Then you can travel and explore a three-dimensional virtual world. While exploring the virtual world, you see other avatars, and they see you. You can talk to them by typing.

Currently, four avatar chats are available online:

- MSN's V-Chat
- CompuServe's WorldsAway
- World's Inc. WorldsChat
- The Palace

Figure 5.4 shows a virtual room in The Palace.

FIGURE 5.4:

A virtual room filled with avatars from the virtual chat world called The Palace

You will find that moving through three-dimensional virtual worlds as a graphical avatar meeting other graphical avatars is a greedy bandwidth application. It's a prime example of the new applications appearing on the Internet and online services that are much more enjoyable when you are able to use a faster connection provided by ISDN services.

Online Information Access Case Studies

The following are case studies of how some educators, businesses, and health professionals use online information access.

Education Online

Now you don't need to leave home to go to school. Educational institutions have realized the potential of providing online courses to allow instructors and students to communicate, regardless of their location.

Berkeley Extension Online

The Alfred P. Sloan Foundation of New York recently granted the University of California at Berkeley $2 million to launch one of the largest educational online projects in the history of the Internet. The Berkeley Extension Online hopes within three years to offer more than 175 college courses that focus on continuing adult education.

Berkeley Extension Online will be developed in collaboration with UC Berkeley Extension's Center for Media and Independent Learning (CMIL). One of the largest continuing education providers in the country, the UC Berkeley Extension represents the continuing education arm of the University of California. Currently, the UC Berkeley Extension offers 25 courses through America Online.

Besides being available to students from around the world, the Berkeley Extension Online project will differ from a traditional classroom structure of lectures and large class sizes. Instructors will place their course materials online, including assignments, and respond to students on a one-to-one basis as they study. Forums and chat groups will be provided for students to access and discuss course-related subjects.

The online offering will include certificate programs in telecommunications engineering, computer information systems, direct-mail marketing, and English for non-native speakers, as well as individual courses in subjects from Visual BASIC and hazardous materials management to interior design and intellectual history. Noncredit and professional or college credit courses will be offered.

Autodesk and University Online

Autodesk, Inc. has announced an agreement with University Online (UOL) to develop a "virtual campus" system to serve as a location where Web users can purchase training software, learning resources, and online courses for Autodesk 2D and 3D design, modeling, and animation software over the Internet. The virtual campus will give Autodesk's business partners, including authorized Autodesk Training Centers, dealers, developers, and publishers, an important new vehicle to reach the customer and to provide content on the system. The virtual campus will be accessible by industry professionals, educators, and students worldwide.

The virtual campus will consist of two primary components:

- The "Classroom" will offer computer-managed training courses on Autodesk software from Autodesk developers, publishers, business partners, as well as the educational community.

- The "Store" will allow Autodesk value-added software developers to advertise and sell their training products and services online.

In addition to providing products and materials for purchase, Autodesk's virtual campus lets users view demonstrations of new products and software releases, as well as participate in software certification and assessment programs.

TIP

University Online's education technology and extensive courseware library provide low-cost, reliable, and convenient "distance" education through the World Wide Web. University Online has also partnered with other major corporations, such as Dun and Bradstreet Inc. For more information, check out the home page (http://www.uol.com) or contact Robert J. Manfredi of University Online at 703-533-7500 or manfredi@uol.com. Autodesk supplies PC- and UNIX-based design software and PC multimedia tools. The company's products are used in many industries for architectural and mechanical design, film and video production, video game development, and Web content development. Its Kinetix division handles PC-based 3D modeling and animation software. For more information about Autodesk, call 415-507-5000, type GO ADESK on CompuServe, or visit its Web site at http://www.autodesk.com. Kinetix can be reached by calling 800-879-4233 or through its Web site at http://www.ktx.com.

Business Online

Bank customers seldom visit the lobby of a bank anymore. Many of us now rely on ATMs for our banking transactions. The banking industry believes that opportunities to cut costs and increase services exist online. One excellent example is Security First Network Bank— no bricks and mortar, this is the first 100 percent virtual bank. You can only visit the bank online! Security First Network Bank is the world's first federally approved, completely online bank, certified by the Office of Thrift Supervision (OTS, which serves as the federal regulatory body for the savings bank industry).

This virtual bank, created by Atlanta-based Five Paces Software Inc., lets users perform most banking transactions via the World Wide Web. Not only can users open accounts and pay bills, but Security First lets them track and itemize their spending. Hyperlinked statements and reports can be pulled up to review expenses by category. By clicking on any item, users can see when their "checks" cleared. Recurring bills can be paid automatically each month.

There is still a lobby, but it's an electronic one, as you can see in Figure 5.5.

FIGURE 5.5:

Security First's electronic online lobby

The Security First site offers browsers a demonstration of their services. Also, the site has plenty of help, an online FAQ (frequently asked questions) button that answers many questions, and a section on banking security on the Internet.

Here are some other examples of doing business online:

- With Federal Express tracking and shipping software, you can track your package, wherever it originated and wherever it is headed. You will know if it has been delivered and who signed for it or if it is still in transit. You will always know the current status of your package in the Federal Express system.

- A government contracts consultant uses the Internet's resources—for sending e-mail and doing research on marketing and government data—to stay competitive with bigger companies with a global presence. Consultant's are very aware of the business axiom "time is money." But even with a high-speed modem, Web browsers are slow. After replacing his analog modem with ISDN service, the consultant found that Internet navigation became six times faster. The speed of ISDN now makes it feasible, both in terms of online costs and billable time, for him to expand his reach to access photos, graphics, illustrations, charts, and diagrams. These capabilities add to the quality of the consultant's presentations, proposals, and other work.

- Financial and insurance companies are using ISDN to speed access to online stock quotes and financial information. Real-time prices from the stock exchange can be downloaded quickly through dedicated links. Traders and investment managers can work from home and perform investment analysis using ISDN links to computers and databases located in their main company location. Teleworkers also access financial information directly from Wall Street via ISDN.

- A radio station uses ISDN to provide high-quality digital links between remote locations and their broadcast center. With five ISDN lines for a digital-audio application, the radio station has reduced its costs and can provide better-quality audio (a wider frequency range). These capabilities allow for more creative programming and give the station the flexibility to broadcast from sites around the country.

Health and Medicine Online

Another area where online access applies is health and medicine. Along with the ability to transfer medical images, online information can help with public safety and patient management.

Medical Teleradiology

The range of images used in medicine includes computerized tomography (CT or CAT scans), ultrasound, nuclear medicine, magnetic resonance imaging, and of course the traditional x-ray. In fact, with the exception of x-rays, all of these originate as digital files, which are most often viewed on a computer screen. Furthermore, many x-rays in specialized applications, although they begin as films, are being scanned into digital files.

The potential for transmitting these files is enormous. For example, medical files could be sent:

- To specialized radiologists for analysis and diagnosis
- To remote experts for consultation
- To surgeons
- To practiced technicians for image enhancement
- To administrators to file with a patient's record

Unfortunately, the files are almost always huge. With ISDN, the tedious and impractical transfer with an analog modem is gone. ISDN, not only saves money in online costs, but it also saves time when time is critical. For example, a radiologist no longer needs to rush to a hospital in pre-dawn hours to look at an accident victim's CT scan. That image can now be transmitted to a PC at the radiologist's home, where he or she can prepare a diagnosis while still in robe and slippers. In addition, hospitals and specialized radiology laboratories are becoming libraries for these images, so that the speed of ISDN can make them available whenever and wherever they are needed.

Public Safety

In Colorado Springs, Colorado, the fire department uses ISDN to support dispatch assistance for firefighters. Dispatchers rely on ISDN's speed and accuracy to retrieve information about which trucks to dispatch and information about hazardous materials that may be near the site of a fire.

At the U.S. Army's Redstone Arsenal in Huntsville, Alabama, ISDN enables emergency dispatchers to be fed information automatically relating to incoming emergency calls. They can discover the location of the calling party and key information linked to the location, such as the presence of munitions or combustible materials.

Medical Patient Data Systems

Pacific Bell's HealthLink and PhysicianLink offer both hospitals and physicians a better way to cope with, access, and use the growing tide of patient information that is often stored in many locations. Increasingly based on the speed and flexibility of ISDN, the system gives doctors and other health-care providers quick and full access to stored data such as:

- Doctor's notes made during a patient's office visit, prescriptions, and the like
- Results from urine, blood, and other tests prepared in a range of locations
- Patient records, surgical proceedings, medicines prescribed and administered, as well as x-rays and other images generated during a hospital stay
- Radiological images generated at specialized laboratories
- Insurance and billing records

The results are a much more streamlined flow of information, better data for physicians, faster and better diagnoses, and more effective and efficient health care.

Online Information Access Tariff Considerations

Your tariffs, or costs, for using ISDN service for online information access applications depend on your ISP or other online service provider, as well as on your phone company's charges. The following sections describe ISP ISDN service, ISP service plans, and fees for other access providers.

ISP ISDN Service

ISPs usually offer two kinds of ISDN Internet service:

- Dial-up service is normally chosen by residential customers. Dial-up service is a temporary connection, established only as needed.

- Dedicated service is a permanent connection, available 24 hours a day.

Usually, companies with LANs in large offices or Web servers require a dedicated 24-hour connection. Both dial-up and dedicated services are configured to support either one B channel (64 kbps) or two B channels (128 kbps). For example, you may choose an ISDN Internet service that offers 50 hours a month for a set fee. However, most Internet service provider's charge by the "B-channel" minute. This means that if you use two B channels to surf the Web at 128 kbps, your monthly allocation of 50 hours will be consumed twice as fast.

ISP Service Plans

ISP service plans tend to fall into two categories:

- You can buy a block of time by the month, quarter, or year. Usually, this type of plan sets a price for a fixed amount of time (hours) each month.

- You can pay for your service by the throughput (the amount of data or bytes) you download monthly. As you may know, each time you travel to a different Web page, the information is downloaded to your computer.

The throughput service plan is usually assigned to business Web pages, whose owners are unable to determine the amount of traffic visiting their Web site each month. However, most ISPs charge by the B-channel minute.

Currently, competition appears to be keeping most ISDN ISPs from charging for the second B channel's minutes. ISDN dial-up lines are often staffed by an Ascend Max 4000 (a router), which supports dual B-channel bonding and dial-on-demand. To promote this feature, ISPs will offer you an additional B channel, giving you the additional bandwidth at no extra charge, if one is available. But, in some cases, the second B channel will not be available. Similar to when you use analog-based services, when you dial up your service provider, you may experience a busy signal. Or, a B channel might not be available when your ISDN equipment requests that second channel, and your bandwidth or speed will be limited to 56 kbps or 64 kbps, depending on your area.

On the average, ISDN Internet services provide high-speed access to the Internet for approximately $1 per hour. For example PSI provides ISDN national Internet services for $29 per month for 29 hours, with each additional hour for $1.50. UUNET is another service that provides ISDN access to the Internet. It charges $195 per month for 50 hours and $2 for each additional hour.

Figure 5.6 shows a comparison of the costs with one and two B-channel configurations and with a 28-kbps analog modem. Using an approximation of $1 per hour, the chart shows the relative costs of downloading two different files, 100 kilobytes and 1 megabyte, using the three configurations.

FIGURE 5.6:

Relative costs of downloading files with one B channel, two B channels, and a 28,000-bps modem

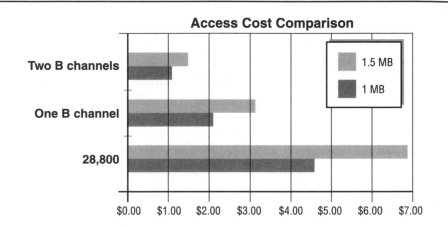

Fees for Commercial Online Access Services

Commercial online access services also offer Internet access in addition to other proprietary services. The number of online service providers continues to grow. Currently, the four main services are America Online, CompuServe, the Microsoft Network, and Prodigy. Table 5.1 shows representative fees for these services.

TABLE 5.1:   Charges for Some Commercial Online Access Services

Online Services	Monthly Fee	Hours	Additional Hours
America Online	$9.95	5	$2.95
CompuServe	$9.95	5	$2.95
The Microsoft Network	$4.95	3	$2.50
Prodigy	$9.95	5	$2.95

Keep in mind that all the prices quoted here are subject to change. Be sure to verify charges from each provider before you make any purchasing decisions.

About Telecommuting Applications

- A definition of telecommuting

- Advantages of telecommuting

- The home office, with and without ISDN

- Corporate telecommuting policy

- Some telecommuting case studies

- Telecommuting tariff concerns

How many times have you seen business cards with phone numbers and addresses that do not match geographically? This is becoming more common with small businesses, sales representatives, and even large corporation employees. I met someone with an office in Irvine, California, and a phone number in Santa Monica, California, which is about 60 miles north.

And how many sales representatives give the corporate regional office as their main address, but really have all their phone, fax, and e-mail addresses based at their home office. This practice is becoming common in more businesses and in more departments or functional groups.

The corporate campus that housed the monolithic companies of the 1970s and 1980s may soon disappear entirely. Downsizing, decentralization, and re-engineering of the work place have increased the use of outside consultants and contractors.

In large geographic areas such as California, residential property values are high, and affordable housing has moved farther and farther down the freeway and beyond the suburbs. Many captive workers want to end long commutes. Many ordinances are directed at reducing or eliminating this commute. The environmentalists and new generation of mobile workers want to ease congestion on the highways and increase their own quality of life. Working at home suddenly seems to benefit both workers and companies.

Today, there are some 40 to 43 million people who are telecommuting in some fashion. The number is even higher if you add the people who are using online services for simple e-mail and other business information exchange. The number of self-employed and corporate employees eligible for their company's telecommuting programs has doubled in the past three years.

Larger data files and graphical information are hitting the limit of the modem technology and also taxing the patience of the users who are sending and receiving these files and this information. As more information and more people create the demand for increased bandwidth, ISDN is becoming more integrated into PC environments and becoming available at retail outlets. This creates the demand for ISDN as the next logical step to working at home. You can communicate with headquarters, remote clients, or distant support resources. Many "work-at-home" types now see the need for quick exchange of the information, and they know that there are options beyond Federal Express and regular analog phone service.

This chapter answers the question "What can I do with ISDN?" with "You can telecommute!" For details on specific requirements and equipment for telecommuting with ISDN, see Chapter 11 in Part 3 of this book.

Who Is a Telecommuter?

A single simple profile does not apply to the "typical" telecommuter. The telecommuter may be using a combination of the applications that are highlighted in this book, or he or she may just be dialing in using standard POTS (Plain Old Telephone Service) to collect e-mail messages remotely.

Telecommuters can be categorized in many different ways, such as:

- Corporate, full-time or part-time
- Mobile worker, full-time or part-time
- Self-employed (or business-at-home entrepreneur), part-time or full-time

Categorizing the telecommuting functions has focused on corporate versus self-employed, infrequent user versus frequent user, or light user versus power user descriptions, further split into full-time or part-time of each segment. Analysts and consultants have conducted many studies that attempt to define these telecommuter segments. The definitions are based on the amount of time the person works at home and the number of other people the worker needs to communicate with for his or her job.

A simple definition of a telecommuter that we can apply here is someone who works at home or from a remote office—someone working away from the traditional office environment. For example, at one time, I was a telecommuter, with an office at headquarters and six remote sales offices in different parts of California. While I was visiting sales teams and customers in California, I telecommuted to my headquarters office.

I have a business associate who has been telecommuting from her home and never really had an office. She is a consultant and finds that all of her meetings are at her client's or another business's office. Full-time telecommuting allows her to avoid the costs of the traditional office with its typical office overhead.

At Pacific Bell, I had another full-time telecommuter associate. She was able to avoid the commute across the San Francisco Bay by working in her home as a product manager. With a fax machine, PC, printer, voice mail, and call features (like call-waiting, call-forwarding, and caller ID), she could do her work and respond to requests as effectively as if she were located in the headquarters office across the Bay. She worked more productively and longer, and could make more flexible arrangements for her daughter's childcare during work time.

Many homemakers today are like part-time telecommuters who are running households and using some of the features of telecommuting for increased productivity. Who says running a household and raising children is not really working at home?

And most corporate workers also must be part-time telecommuters to keep up with business while they are not at the office. Very few full- or part-time white-collar workers can survive without some form of telecommuting or work-at-home activity.

About one-third of all households have a need for some form of telecommuting; of those households, two-thirds have second lines and a personal computer in their home. These homes also tend to be the ones that have the highest usage of phone service today. It is projected that by the year 2000, telecommuting may reach two-thirds of all households.

Why Telecommute?

Whether you're a potential entrepreneur deciding whether to run your own business from your apartment or an MIS manager organizing your department, telecommuting is an option you should consider. The following sections summarize the benefits for corporations, employees, customers, small business, and the community.

Corporations and Telecommuting

For corporations, telecommuting can have benefits such as these:

- **Increased productivity:** Telecommuting empowers employees to work more effectively. For example, the State of California pilot program found that telecommuters accomplished 20 percent more than office workers.

- **Improved employee recruitment and retention:** Telecommuting allows you to broaden the geographic base for company recruitment. You can employ contractors and flex-time workers and reduce turnover and relocation problems.

- **Reduced costs:** When workers telecommute, you can cut physical plant and maintenance costs. You will expend fewer resources producing, copying, faxing, and delivering printed information.

- **Strategic flexibility:** Telecommuting lets you move quickly to seize market opportunities, tap expert resources, and pursue partnership ventures without regard to location. You can put information and talent out on the "front line" close to customers.

Telecommuting Employees

Employees also benefit from telecommuting:

- **Time savings:** Telecommuting reduces commuting time, or eliminates it completely. You can accomplish more in less time.

- **Balance of work and family obligations:** As a telecommuter, you can choose flexible work hours and more easily accommodate your children's sick days and vacations.

- **Higher quality of life:** Telecommuting can reduce stress. You won't need to deal with rush-hour traffic, and you can work at an efficient, relaxed pace in a calmer environment. You'll also be able to spend more time with your family.

- **Increased productivity:** It's easier to concentrate without the distractions and interruptions of a traditional office, and when you can work any time of day or night.

Customers and Telecommuting

Customers like telecommuting for reasons like these:

- **Convenience:** You can do business at local branches or remote sites close to your customers.

- **Better service:** You'll be able to interact with company personnel who understand the community and are attuned to local customer needs.

- **Increased availability:** Given the appropriate telecommuting setup and equipment (including a second line, fax machine, voice mail, and a separate work space), most people can be more accessible and more available for customers and clients.

Telecommuting for Small Businesses

For small businesses and entrepreneurs, telecommuting has these benefits:

- **Overcome typical small business limitations:** The size of your business will not restrict the reach of your business and clients. Your physical location will not limit business. You can communicate and do business on a national, or even worldwide, scale.

- **Access to information:** While you're telecommuting, you will have access to vast resources of online information, from the Internet and other online services.

- **Better customer support:** You will be more available to your customers when they can communicate with you remotely "around the clock."

TIP Telecommuting is a good option for small startup companies. Rather than starting a business with a large office overhead, the company can find a very small office for the few principals, and everyone else can work from their own home or regional small office. This saves money and allows you to cover a variety of locations across the country or region.

Benefits of Telecommuting for Communities

Telecommuting even has benefits for the community as a whole:

- **Improved natural environment:** Less commuting means less traffic, which means reduced air pollution.

- **Improved man-made environment:** Telecommuting allows workers to move away from heavily populated areas. Reducing urban congestion can expand the activity in and increase the vibrancy of suburban "bedroom communities" and rural areas.

- **Economic boost for rural areas:** Corporate workers and highly skilled professionals can now live outside densely populated cities. Many rural areas around the country are seeing an influx of professionals escaping the major metropolitan areas, seeking a higher quality of life.

Telecommuting Buzzwords Brief

Here are some terms you may come across in discussions of telecommuting:

- *Connect time* is the time it takes to initiate a call, dial the phone number, have the remote location respond to your call and physically connect to your line, and start sending information or data.

- A *corporate telecommuter* is a person who is employed full time by a company or corporation, and works full time, part time, or after hours from an office at home.

- *E-mail* is electronic mail, or the messages you can exchange through your personal computer and online services or the Internet.

- *MOHO*, for Mobile Office Home Office, refers to the group of mobile workers who travel to various sites.

- *Multimedia* describes information or files that include text, video, and/or audio. Many World Wide Web pages include text and animation, which use multimedia data transfer.

- *RLA*, or *remote LAN access*, is an application similar to telecommuting, except that it adds connections to LANs from remote locations. Chapter 8 provides details about RLA applications.

- *ROHO*, for Remote Office Home Office, refers to the group of workers who use RLA applications from home.

- *SOHO*, for Small Office Home Office, refers to the group of workers who are based in small home offices.

- *WAH*, for Work at Home, is a broad category that encompasses all the people who work at home.

Telecommuting Design Considerations

For your telecommuting setup, you need to consider several design issues:

- What is your home office arrangement or work space?

- What equipment do you need to perform your job or run your business? What equipment do you have already that may be used for your home office and communications?

- Can your house or living unit have more than one phone line?

- Is there a corporate or company policy? Are company standards and support already available? Do you need to create any standards?

- How many hours will you be working at home?

- What features or services will you need at home?

Today's Home Office with Analog Telephone Services

Conventional analog telephone services don't satisfy many of today's home office users, because data transmission rates are too slow, application standards are not well established, and integration of voice, data, and fax calls is inadequate. ISDN is available for fixing the weak link in the mobile, home office. Before exploring ISDN in detail, you should understand how most home offices are configured now.

With powerful and more cost-effective computing equipment, workers can be as productive at home as in the corporate office. But the home office has one serious weakness: lack of network communications equipment and services. In addition, many workers are trying to benchmark the communications capabilities that they have at an office with the network communications they may have at home. Since they have had the comforts of an office at a traditional location, they hold the network communications at this office as their standard. When they work at home, workers may expect the ability to connect to the LAN, send and receive faxes, and handle multiple phone calls at once. This may or may not be what can be in a home office; the workers' capabilities depend on the number of lines and equipment in their office.

Today's home office uses conventional analog telephone circuits, which are fine for voice calls, but too slow for accessing e-mail from an online service, getting technical data from the Internet, and sharing files with co-workers. It's nearly impossible to use multimedia or videoconferencing applications at such slow network speeds.

> **NOTE**
>
> Developers are constantly trying to extend the life of the analog modem technology with and without the Internet. For example, Intel recently announced support for videoconferencing on an analog line. But most of these developments fall short of the two-line capability of ISDN service and lack the quality of videoconferencing over ISDN connections. We'll take a look at videoconferencing applications in Chapter 7.

Your home office may have a phone, fax, modem, and computer for communication with remote locations like client offices, online services, and Internet service providers. The traditional public-switched phone network connects the home office to all these locations and information sources. Figure 6.1 shows examples of standard configurations of equipment in a home office.

FIGURE 6.1:

A home office may contain several types of communications equipment.

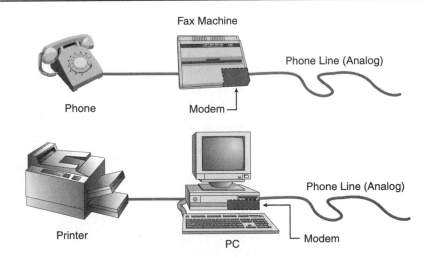

Your setup may include the modem functionality integrated (or included) inside your computer or fax machine. You might also have a computer with a fax card that allows you to receive and send faxes through your computer. If you have two phone lines, you can call someone or receive calls while you're faxing or receiving faxes, or receiving information from an online source. Two lines allow you to be more accessible for phone calls and fax transmissions.

NOTE Most fax machine manufacturers and resellers recommend that their customers install separate dedicated lines for their fax machines. So most home offices have a phone line that is used infrequently. A partial solution is to have the phone and fax share a line. You can buy a device that automatically detects incoming fax calls by their calling tone and tells your fax to answer the call instead of your telephone. If the calling fax sends the correct tone, the device works fine. Unfortunately, there aren't any devices designed to automatically identify a modem (data) call, so your modem still needs a separate line.

With this setup, you need to pay for individual lines for the phone, fax, and modem—if you can actually get three phone lines to your home. The telephone wiring in most average houses is only capable of handling two phone lines. Of course, if this is all you need, then you can stick with POTS for awhile.

WARNING Your home's wiring may impact the installation of your ISDN service. Many homes are wired for two pairs of wires, which allows for two analog phone lines. However, if a home is wired in series—one phone jack is wired from the first location to the second, then third, and so on—you may run into crosstalk problems. This means that one phone conversation can be heard on the other line. In some cases, you may need to have your ISDN phone jack wired into your house or home office separately from all your other analog service.

Tomorrow's Home Office with ISDN Telephone Service

The ISDN-connected home office uses ISDN for data communications and normal telephone services for voice. ISDN greatly improves the speed and cost of accessing e-mail from an online service, getting technical data from the Internet, and sharing files with co-workers. Multimedia technology, such as videoconferencing, is also possible. The system components and the connections are similar to the analog counterpart, with one important exception: ISDN telephone service.

If you decide to use ISDN services, you will need to purchase a new piece of equipment for your office, an ISDN terminal adapter (TA). This is a modem-like box that can support up to 128 kbps data—four times the rate of the fastest modem available. Figure 6.2 shows an example of a home office equipped for ISDN services.

FIGURE 6.2:

A home office with an ISDN line and terminal adapter (TA), with two analog conversion jacks

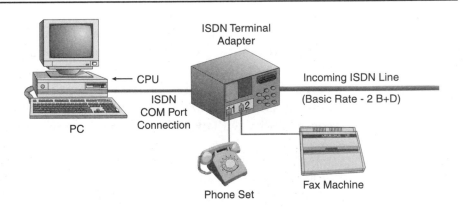

You'll also need to keep your modem if you want to access services designed for analog modem communications. A frequently asked question is "Why do I still need a modem when I have ISDN service for data calls?" Unless you have a modem, the ISDN-connected

home office won't be able to communicate with many established modem sites, because ISDN data calls are incompatible with modem data calls. Common terminal adapter equipment has no provision for simulating a modem and cannot exchange digital data using voice-call connections. To maintain compatibility with the currently installed base of modems, you will need a modem.

Having two different solutions for data transmission means that incoming data calls must arrive at the correct phone number for each type of call: the ISDN number for ISDN data calls or the analog number for modem data calls. It's difficult to match the incoming calls to the correct phone number. Fortunately, the telecommunications industry is working to expand the functionality of National ISDN standards so that terminal adapter vendors can include an integrated modem function along with the ISDN network connection in future equipment revisions. The next evolution of some of the terminal adapter equipment will provide this feature, but it isn't widely available yet.

A low-cost solution is to purchase a terminal adapter with a special analog port that can connect to the modem equipment you probably already have. To place a modem call, simply make a modem connection over a voice call, the way you've done it for years, only now the data goes over an ISDN voice call. The advantage of this approach is compatibility with existing modems at a low cost.

With this setup, the modem connects to the analog port, an un-labeled box on top of the terminal adapter diagram. That analog port simulates a central office analog phone line, so an external modem thinks it is connected to a regular phone outlet. In fact, the analog ports are useful for connecting analog phones to the ISDN network also. Digital phones are expensive, and much of the digital phone functionality is available through new computer telephone integration (CTI).

Is There a Corporate or Company Policy?

If you are working in a business environment, you will need to discover if your company has specific policies regarding telecommuting. Without a policy, it will be nearly impossible to provide or receive any common resources or support for the successful implementation of any telecommuting program—with or without ISDN as the network service.

If you are the MIS staff, you will find that *you* are the one responsible for setting the company policy and standards for telecommuting. Not only do these guidelines need to be developed, but they also must be approved and supported at the highest level.

Your company's telecommuting policy should cover the following:

- The funding required for a telecommuting program
- The level of executive support, or sponsorship from your company, for telecommuting
- The security of company access and the security requirements
- The employees who will have access to the telecommuting program
- Standard equipment and services that will be supported, including the standard configurations (for example, PC requirements, including memory and communications hardware)
- Standard agreements for equipment purchasing and maintenance, including repair policy and provisions for spare equipment
- Service levels and hours of support, including on-site service at home (if provided)
- Provisions for remote LAN access, if allowed
- Training or employee services

- Time-reporting requirements (for tracking the number of employees and the number of hours they telecommute)

- Employee contract for telecommuting (to ensure appropriate work activity and extend company policy into the home office)

Much of this policy does not deal with technical details, but with business and political decisions of your company. Each company and employee needs to judge by position, project, and function the productivity gains and benefits for performing this assignment while part- or full-time telecommuting. The program should not be strangled by a strict corporate policy or abused and used simply for the employees' personal convenience.

In many cases, you will find that only certain management levels, work groups, functions, or job titles are offered telecommuting programs. These requirements must be defined and then administered.

Also, new management techniques are required for managers to work with and gauge the productivity of telecommuting subordinates. This is not an easy transition for many traditional managers.

Most companies offer telecommuting as a general option and leave it to the local management's discretion to agree to telecommuting and make arrangements with telecommuting employees. For example, one company defined three variations of telecommuting and gave the local management the authority to approve one of the three: infrequent, frequent, or full-time telecommuter. Each one involved the same basic agreement, but the infrequent and frequent telecommuter received less equipment and support from the company. The premise was that a full-time telecommuter uses the home office as the sole work place, so it should be equipped as a full office. An infrequent telecommuter still uses the traditional office and requires all the same hardware and services as a full-time, regular employee. What this policy means to the telecommuters is that you need to be a full-time telecommuter to get a printer and fax machine in your home office, but you also no longer have permanent office space in your regular building.

Your company's telecommuting policy is also governed by the executive support for the program. Some do not really sponsor telecommuting, so the equipment and service support is weak (or nonexistent). On the other hand, many organizations strongly support telecommuting and allow complete flexibility in the program. They allow the work groups and employees to decide how they can work best. You'll find that a telecommuting program's success depends on the executive or officer level support for telecommuting and where the managers see telecommuting within their respective companies. If a company does not sponsor the telecommuting option strongly, it is usually not exercised.

Some Guidelines for Developing Your Telecommuting Policy

If you have a small company, with say 250 employees, your telecommuting policy should be short and specific to the work group and the standard equipment that you will supply. The telecommuting policy could be as simple as getting the supervisor's approval. However, you still will need to define the work groups that are eligible and outline a few standard equipment configurations. You also will need to decide the level of access to company resources, such as LAN access, faxing locations, and so on, that telecommuting employees will have. In smaller companies, you may need to depend on the employees to manage and arrange services or equipment at their homes, with limited equipment support provided by a local equipment provider.

If your company is larger, with maybe 8000 employees, your telecommuting program needs to be more detailed. You will probably need to allow for more standard configurations. It is likely that you will have PC and Macintosh systems, and the telecommuters will want access to a much more varied list of company resources and databases. The policy

(continued on next page)

could be managed by local management but supported by the information systems or MIS departments to ensure that you can run your mission-critical business applications and systems while allowing for telecommuting and remote access.

In most cases, the first decisions after gaining management support for telecommuting are:

- Who is eligible?
- What access do they need?
- What equipment will we support?
- What level of support will be provided by the company? At what hours?
- What is the technical sophistication level of our prospective users?

Typically, when companies embark on creating this new policy, they find that the success of their small-scale trials depends on the selection of their trial participants. If your trial participants really want to make the program work, and also have some of the standard equipment already at home, you are more likely to have a successful outcome. It also helps if the employees are patient and flexible.

Another factor will be the technical sophistication level of the employees who are telecommuting. If they are not that software savvy or PC literate, it is a bigger leap for them to add the network communication layer to their work-at-home office. When you are considering someone for the telecommuting program, along with the work assignment or function that an employee fulfills, also assess the personality of the employee. Some employees need more structure and supervision than others. In addition, if an employee is a poor performer and/or needs close supervision, telecommuting is just a symptom of a bigger problem.

One of the difficulties in developing a telecommuting program is finding the balance of cost-effective programs that allow employees sufficient access and equipment without sacrificing the company's system integrity

You are faced with the increasing costs of supporting too many varied equipment configurations options. Not many vendors or service companies are anxious to service equipment in the home, especially PC equipment and peripherals that can vary so much. To avoid high costs for equipment, provide a few standard configurations of equipment and service levels that satisfy the majority of your employees or clients. Support for every configuration and every network connecting device will dramatically increase the costs of a telecommuting program.

These factors apply with or without ISDN service as your telecommunications choice. However, ISDN service does involve some additional support and slightly more complex equipment options. Be aware of the basic ones in addition to the ones added by ISDN service. Again, standardization of what is supplied or supported is the key to the success of a telecommuting program.

Who Is a Candidate for ISDN Telecommuting?

One of the first things that must be done for any ISDN telecommuting program is to classify the telecommuters. This definition should be part of your company's telecommuting policy, as described in the previous section.

Of this total number of telecommuters, some may not have any needs beyond basic telephony—voice communications—to do their business. Those users don't need data connections or faxing facilities. However, you'll probably find that these telecommuters are a minority. Most workers need to communicate electronically through some form of online access.

Today, in the United States, the majority of telecommuters are using regular analog dial service. Although we have heard of cases of some power telecommuters (usually subsidized by their companies) using full T1 (24 channels) or frame relay service, roughly 90 percent of all telecommuters are using regular phone service. ISDN users represent less than 3 percent of all telecommuters now, but this is projected to grow to at least 30 percent, and as high as 60 percent, by the year 2000. This 30 percent could represent as many as 15 million ISDN users.

Telecommuting does not require ISDN service, and ISDN did not create the trend for people to telecommute. There will also be a large base of telecommuters who will be satisfied with analog phone service. For the infrequent telecommuter or very-low-bandwidth application, analog phone service is a good, cost-effective network solution. But ISDN service does greatly enhance telecommuting, and telecommuters who have tried ISDN prefer it because of its increased reliability and quicker transfer speeds.

There are three critical factors for an ISDN telecommuter:

- Frequent use of telecommunications
- Large amounts of data exchanged
- Simultaneous use of the line (for fax, voice, and/or data applications)

These three factors identify the need for the faster connections and increased data-transfer speed provided by ISDN service. Figure 6.3 illustrates the differences between users who need ISDN services and those who don't.

Who qualifies for ISDN
telecommuting?

Telecommuter

- Frequent use of telecommunications services
- Large amounts of files or data exchanged
- Simultaneous use of phone line for fax, voice calls, or data applications
- Need to connect to remote systems or online services
- Home office requirement for personal needs or work position

Non-Telecommuter

- Limited need for work at home environment
- Never needs access to office or business information
- PC not required at home for work
- Work activity rarely required after hours

Telecommuting Case Studies

The telecommuting trend has led to a demand for more frequent connections and increased bandwidth. It may have started with simple one-page faxes or a few e-mail exchanges, but now telecommuting involves exchanging a whole lot of e-mail and transferring big documents. The increase in fax communications and transactions is another part of the growth in telecommuting. And, of course, the number of personal computers in the home is another indicator of the full- or part-time work-at-home trend. Let's take a took at some actual examples of ISDN being used for telecommuting applications today.

3Com Increases Employee Efficiency

The need to move data and information are not confined to the office walls. An ever-increasing number of businesses today have a noticeable percentage of employees that telecommute. These employees must have the same access to other employees and information as they would if they were seated among their counterparts at the office.

3Com is one of the companies leading the way in the telecommuting trend. Pacific Bell's ISDN services and the 3Com Impact ISDN digital modem have allowed 3Com to maintain a comparable level of access for all of their telecommuting employees. 3Com began using Pacific Bell's ISDN Basic Rate Service in its telecommuting program in November 1995. Currently, approximately 250 3Com employees are established in home offices, and this number may quadruple by the end of the year.

The telecommuter uses the 3Com Impact ISDN digital modem with a Basic Rate ISDN line to dial into a Primary Rate ISDN line at the office location. This provides the user end-to-end digital connectivity at speeds that virtually replicate his or her setup at the office. The telecommuter can function in the same manner to gain access to information and other individuals.

The Olympics in Atlanta

ISDN went prime time in practical use for the summer Olympics in Atlanta in 1996. Bell South filed a special ISDN tariff just for the Olympics.

The primary application for these lines was for the journalists and press to send their stories to many locations worldwide during the Olympics. Although there was some amount of videoconferencing use for these ISDN lines, the reporters primarily used them for transferring, exchanging, and updating their news stories from their temporary home base in Atlanta.

This is an ideal telecommuting application for ISDN service in a real, demanding environment. It demonstrates the power of enhancing the speed of data communications using ISDN on a worldwide scale.

Pacific Bell Promotes Telecommuting

Pacific Bell has been one of the early sponsors of telecommuting. Its policy was originally somewhat loosely described for employees, but has now been updated with three tiers, or types, of programs for telecommuting employees. The program is well-documented and provides enough variety to address most organizations and employees. Because of the size of the company, with its 50,000 plus employee body, the multi-tiered program has a limited amount of flexibility and uses very standardized equipment.

ISDN is an official option of Pacific Bell's telecommuting program. With the increase in the number of employees and applications, the need for ISDN to be part of the program was inevitable. And Pacific Bell also had the "use-what-you-sell" motivation. But the need for more bandwidth and the increased frequency of telecommuting were the key factors that prompted the demand for ISDN to be part of the program.

The members of the information systems group were the first power users to have ISDN service, but the sales and the marketing groups quickly followed. E-mail use was outpaced by Internet access, home page research, and file transfer. Using these applications increased the file-transfer size by a factor of ten to a hundred times the size of the standard e-mail message.

TIP

You will find some valuable telecommuting information and background on Pacific Bell's Telecommuting Resources Guidelines page on the World Wide Web. It is in the Products section at: www.pacbell.com/. The home page consolidates a wide variety of information from around the world, and from many different perspectives, on the subjects of telecommuting, teleworking, and flexible work.

Intel Starts ISDN Programs

During 1994, Intel introduced an employee telecommuting program at its Santa Clara, California, facility. Now there are more than 1000 employees who telecommute using ISDN lines at that facility alone, and the program has expanded beyond that location. ISDN was the service of choice because it allows LAN access and simultaneous use of the line.

The enabling technology for these Intel workers was Intel's Proshare product line, which allowed them to communicate in a multimedia mode while at home. The use of videoconferencing and desktop collaboration at the same time created the interest in ISDN as the network communication option. Employees in Santa Clara and Portland worked on the same project and/or document from their separate homes or offices. Although this setup greatly reduced travel within the company, it did not eliminate travel altogether, just as room-system videoconferencing centers do not eliminate all inter-company travel.

Hewlett-Packard Goes Mobile

One of the largest national implementations of telecommuting with or without ISDN service on a national basis is Hewlett-Packard's program for employees in all of their major offices in the United States. In keeping with its reputation for work places with a positive atmosphere, and as an extension of the "HP Way," telecommuting has been embraced by Hewlett-Packard for many years.

In late 1994, Hewlett-Packard began to test ISDN service as an option. This service is now offered in all of the company's offices across the country. Hewlett-Packard's program has created more of a demand for ISDN availability, since some of its locations are in suburban or rural areas. Silicon Valley was never a problem, but it was difficult to provide ISDN service to some rural parts of California in the early days of the program.

As with Pacific Bell's telecommuting program, the large scale of employee demand and the respective applications drove the need for ISDN as an option for Hewlett-Packard. The company developed a small set of standard configurations to provide for lower costs and economic service options for work-at-home employees. Some of this standardization meant no Macintosh support and PC support for only some implementations.

Telecommuting Tariff Considerations

The main ISDN tariff considerations for telecommuting have to do with the fact that telecommuting applications usually involve home or residential use. Availability is a factor for any person or company considering ISDN as an option. You will need to verify that you have ISDN service in the residential areas of the telecommuters.

Also, you many want to check for any distance problems or additional costs that are based on your local carrier's tariffs. Many telecommuters find that communicating distances beyond 18,000 feet results in either some transmission problems or increased costs for extended ISDN service.

Most telecommuters should consider choosing a packaged usage plan. Many carriers are now packaging 50 to 60 hours for one rate, which fits the profile of most uses for telecommuting.

Lastly, telecommuter education is important. Idle connections are wasted use of the remote resources; the usage costs increase with little or no productivity. You, or your employees, will need to learn how to use connect time wisely.

About Videoconferencing Applications

- Videoconferencing defined

- How videoconferencing works using ISDN

- Videoconferencing terminology

- Design considerations for planning a videoconferencing system

- Types of videoconferencing equipment

- Videoconferencing applications for business, dataconferencing, and education

- Some videoconferencing case studies

Videoconferencing is not just for "big" companies anymore. Traditionally, communication has always been limited by bandwidth, or the amount of information exchanged between two people or groups (as with smoke signals and Morse code). ISDN is breaking through the "bandwidth barrier" by offering higher bandwidth rates at lower costs.

Interactive multimedia videoconferencing provides capabilities for interaction and collaboration that were previously available only in face-to-face meetings. This type of communication and information sharing will soon be common in all areas of industry and business, government, education, and entertainment.

What Is Videoconferencing?

Videoconferencing allows groups or individuals in different locations to hold interactive meetings and collaborate. In the most basic application, videoconferencing can be limited to a "talking head," or an exchange of images and voices from another location. Usually, the video portion is presented on a television-like monitor or computer screen. With the proper equipment, the collaborators can hear each other while they share video images of each other. This video "window" can be as large as a television screen or as small as a postage stamp. Often, the size and quality of the window is determined by the bandwidth (or the size of the logical pipe) that has been established between the participants.

An enhancement of this basic videoconferencing is *interactive multimedia* videoconferencing. In addition to "talking heads" or video shown on a screen, participants in the interactive multimedia videoconference can share images, files, and even an Internet connection with another individual in another place that has no images, files, or Internet access connection. Besides just passing files, participants can mark up a presentation, change numbers on a spreadsheet, or add comments to a letter that may reside on another computer in another location. Even though the file may physically exist on a personal computer miles away, the file will appear to be on the other person's machine. This exchange is often managed using a "master" and "slave" relationship. One participant owns or has possession of the file as the master and shares the file with the slave. In the simplest of arrangements, PCs at each end can be used to share files or let participants work concurrently on a single computer application.

How Does Videoconferencing Work?

Videoconferencing creates an environment where video and data can be shared between participating locations, interactively. Video requires a huge amount of data (remember "a picture is worth 10,000 words"). In the digital world, one small image takes lots of space. Now imagine 24 pictures (movies show pictures at the speed of 24 frames per second; television at approximately 30) passing by on your screen every second! You get the picture (grin). Plain and simple, video requires *big* bandwidth and requires a big pipe to transmit data from location to location.

As we've said, ISDN offers bunches of B channels, or pipes, to carry the data. However, the bigger the pipe (or number of pipes), the bigger the bandwidth, and of course, the cost.

How can you reduce costs? You can reduce the amount of data by compressing the video—less bandwidth, less cost. However, compression presents a trade-off in quality versus cost. When you compress the video, you reduce the amount of data and lower transmission costs, but the image quality suffers.

The software or hardware that performs this *co*mpression (and *de*compression) is called a *codec*. Codecs are an integral part of any videoconferencing system that uses compression technology.

Why Have Videoconferences?

Using videoconferencing as a stand-alone application, the technology provides an audio and video connection with another location. However, the real pay-off occurs when you choose videoconferencing system components that share many common elements (hardware and software) with your computers and networks.

For example, a videoconferencing architecture based on an open architecture, such as a PC, suddenly becomes a conduit of communication in your organization. The same system can wear many hats. Now, holding meetings, performing a presentation, faxing, and document sharing come together, integrated in a single system.

NOTE　An *open architecture* is one that is not proprietary, allowing third-party developers to legally develop products using specifications that exist in the public domain.

Here are some of the things you can do in a videoconference:

• Participants at all locations can add to or change a computer document.

- You can capture documents, files, and presentation materials as digital "slides" that can be exchanged, marked on, stored, shared, and retrieved from local or remote drives.

- You can create presentations on one PC and save them to any drive on the network or view them on any interactive multimedia videoconferencing PC in the network.

Coupling videoconferencing technology with the flexibility and bandwidth capacity of ISDN allows an office or organization to dramatically leverage its technological investments. With a minimal investment, an ordinary PC can be upgraded to a desktop, personal conferencing system, enhancing the functionality of the PC and improving the productivity and the functional integration of your staff. You get a business solution that is greater than the sum of the individual components. Better yet, the components deliver tremendous value, without requiring extensive modifications to your network, your computers, or your work environment.

Videoconferencing Buzzwords Brief

Here are some terms you may come across in discussions of videoconferencing:

- *Application sharing* is a feature that allows two people to work together when one of the individuals doesn't have the same application or the same version of the application. In application sharing, one user launches the application and it runs simultaneously on both PCs. Both users can input information and otherwise control the application using the keyboard and mouse. Although it appears that the application is running on both PCs, it actually is running on only one. Users can easily transfer files associated with the application, so the results of the collaboration are available to both users immediately. The person who launched the application also has the option of locking the application to prevent the other user from making changes.

- A *bridge*, in videoconferencing vernacular, is a device that connects three or more conference sites so that they can simultaneously communicate. Bridges are often called *multipoint conferencing units*, or *MCUs*. In IEEE 802 parlance, a bridge is a device that interconnects LANs or LAN segments at the Data-Link layer of the OSI model to extend the LAN environment physically. See Chapter 3 for more information about LAN bridges.

- A *codec* (for coder/decoder) is the software or hardware that compresses and decompresses video data. The codec also acts as an interface. Video and audio are run through the codec, which transmits a single, digital signal over a network to the remote location(s).

- *Codec conversion* refers to the back-to-back transfer of an analog signal from one codec into another codec in order to convert from one proprietary coding scheme to another. The analog signal, instead of being displayed to a monitor, is delivered to the dissimilar codec, where it is redigitized, compressed, and passed to the receiving end. This is a bidirectional process. Conversion service is offered by carriers such as AT&T, MCI, and Sprint.

- *Common Intermediate Format* (CIF) is an international standard for video display formats. The QCIF format, which employs half the CIF spatial resolution in both horizontal and vertical directions, is the mandatory H.261 format (a standard for codecs). QCIF is used for most desktop videoconferencing applications where head-and-shoulder pictures are sent from desk to desk. QCIF displays 176 pixels grouped in 144 noninterlaced luminance lines.

- *Compression* is the process of reducing the information content of a signal so that it occupies less space on a transmission channel or storage device. This is a requirement for most video communications. An uncompressed NTSC signal (the standard in the United States) requires about 90 Mbps of throughput, greatly exceeding the speed of all but the fastest and shortest of today's networks. Video information can be compressed by reducing the quality (sending fewer frames in a second or displaying the information in a smaller window) or by eliminating redundancy.

- *Distance-learning* applications incorporate video and audio technologies into the educational process so that students can attend classes and training sessions in a location other than the one where the course is being presented. Distance-learning systems are usually interactive. They are particularly valuable when the students are widely dispersed in remote locations or when the instructor cannot travel to the student's site.

- An *echo* is the reflection of sound waves that occurs when they bounce off an object such as a window or wall. Reflected signals sound like a distorted and attenuated version of the talker's speech. Echoes in telephone and videoconferencing applications are caused by impedance mismatches, or points where energy levels are not equal. In a four-wire to two-wire connection, the voice signal moving along the four-wire section has more energy than the two-wire section can absorb, thus the excess energy bounces back and is returned along the four-wire path.

- *Echo cancellation* uses a mathematical process to "guess" at an echo and remove that portion of the signal from an audio waveform to eliminate an acoustical echo.

- *Echo suppression* is used to reduce annoying echoes in the audio portion of a videoconference. An echo suppresser is a voice-activated "on/off" switch that is connected to the four-wire side of a circuit. It silences all sound when it is on by temporarily deadening the communication link in one direction. Unfortunately, the remote end's new speech is stopped along with the echo, which results in clipping.

- A *frame store* is a system capable of storing complete frames of video information in digital form. This system is used for conversion between television standards, computer applications incorporating graphics, video walls, and various video production and editing systems.

- *Full-motion video*, also known as *continuous-motion video*, refers to video reproduction at 30 frames per second (fps) for NTSC signals (United States standard) or 25 fps for PAL signals (European standard).

> **WARNING** In the videoconferencing world, the term "full-motion video" is often used but often misunderstood. Videoconferencing systems cannot provide 30 fps for all resolutions at all times, nor is that rate always needed for a high-quality, satisfying video image. Picture quality must sometimes be sacrificed to achieve interactive visual communication across the phone network economically. Videoconferencing vendors often use full-motion video to refer to any system that isn't still-frame. Most videoconferencing systems today run 10 to 15 fps at 112 kbps.

- *Handshaking* refers to the electrical exchange of predetermined signals by devices to set up a connection. After handshaking is completed, the transmission begins. In video communications, codecs use handshaking to find a common algorithm.

- *Interactive multimedia systems* allow users to share images, files, and online connections. For example, in an interactive multimedia videoconference, participants can share graphics, spreadsheets, word-processing documents, and database information.

- The *NTSC* (National Television Systems Committee) is the United States standards organization. The NTSC created the eponymous analog broadcast television transmission standard now use in the United States, as well as in Japan and a number of other countries. The NTSC standard is based on a 525-line image displayed at 30 fps.

- The *PAL* (Phase Alternate Line) standard is the European analog color television standard. It uses 625 scan lines per frame, 25 fps, and 2:1 interlace. It is the analog television standard used in

Britain, parts of Europe, and parts of the Far East (with the notable exception of Japan).

- *Real-time* refers to the processing of information that returns a result so rapidly that the interaction appears to be instantaneous. Phone calls and videoconferencing are examples of real-time applications. Real-time information not only needs to be processed almost instantaneously, but it also must arrive in the exact order it's sent. A delay between parts of a word, or the transmission of video frames out of sequence, makes the communication unintelligible. The phone network is designed for real-time communication.

- *RGB* stands for red, green, blue. It is the additive used in color video systems. Color television signals are oriented as three separate pictures: red, green, and blue. Typically, they are merged together as a composite signal; however, for maximum quality and in computer applications, the signals are segregated.

- *SECAM* (Sequential Couleur Avec Memoire) is the color television system offering 625 scan lines and 25 interlaced frames per second. It was developed after NTSC and PAL and is used in France, the former Soviet Union, the former Eastern bloc countries, and parts of the Middle East. Two versions of SECAM exist: horizontal SECAM and vertical SECAM.

- *Switched 56* service allows customers to dial up and transmit digital information at speeds up to 56,000 bits per second in much the same way that they dial up an analog telephone call. The service is billed like a voice line, based on a monthly charge plus a cost for each minute of usage. Nearly all carriers offer switched 56 service, and any switched 56 offering can connect with any other offering, regardless of which carrier provides the service.

- *Talking head* is a term used to describe the portion of a person that can be seen in the typical business-meeting style videoconference—the head and shoulders. This type of image is fairly easy to capture with compressed video because there is very little motion.

- *Whiteboarding* is a term used to describe the placement of shared documents on an on-screen shared notebook, or *whiteboard*. Desktop videoconferencing software includes "snapshot" tools that enable you to capture entire windows or portions of windows and place them on the whiteboard. You can also use familiar Windows operations (cut and paste) to put snapshots on the whiteboard. You work with familiar tools to mark up the electronic whiteboard, much as you do with a traditional wall-mounted board.

Videoconferencing Design Considerations

When deciding on a videoconferencing strategy, you will need to determine what resources your organization has now and what you'll need to get the most out of videoconferencing. For your videoconferencing setup, you need to consider several design issues:

- Interoperability between different video conferencing systems
- The quality of the digital video images
- Communications lines and services for videoconferencing
- The type of videoconferencing system

These topics are discussed in the following sections.

Videoconferencing Standards and Interoperability

As often happens with ISDN, when you choose a videoconferencing architecture or platform for your organization, you need to consider the interoperability of the product. In other words, you need to know if the technology is based on a proprietary standard, and thus limited to products from one manufacturer. Fortunately, to prevent the videoconferencing industry from fragmenting into dozens of "islands" that can't communicate or share with each other, several international standards organizations were created to establish support and oversee standards among vendors throughout the world.

Videoconferencing relies on algorithms and mathematical formulas that determine how data, video, audio, and computer files are compressed, transmitted, and decompressed (through a codec) at the remote end for playback. Codecs use a variety of methods to compress a video signal. An internationally accepted standard for codecs is called H.320 (said H-dot-three-twenty). H.320 defines how codecs from different vendors communicate with each other. H.324 is a new standard for multimedia communications. The standard for audio and high-resolution graphics transfers is T.120. Table 7.1 summarizes video teleconferencing standards.

TABLE 7.1:   Video Teleconferencing Standards

Media	Narrowband VTC (H.320)	Low Bitrate VTC (H.324)	Iso-Ethernet VTC (H.322)	Ethernet VTC (H.323)	ATM VTC (H.321)	High Res ATM VTC (H.310)
Video	H.261	H.261	H.261	H.261	H.261	MPEG-2
		H.263		H.263		H.261
Audio	G.711	G.728	G.723	G.728	G.711	MPEG-1
	G.722		G.711	G.771	G.722	MPEG-2
			G.722	G.722	G.728	G.7xx
			G.723	G.728		
Data	T.129	T.120	T.120	T.120	T.120	T.120
		T.434				
		T.84				

TABLE 7.1 :   Video Teleconferencing Standards (continued)

Media	Narrowband VTC (H.320)	Low Bitrate VTC (H.324)	Iso-Ethernet VTC (H.322)	Ethernet VTC (H.323)	ATM VTC (H.321)	High Res ATM VTC (H.310)
Multiplex	H.221	H.223	H.221	H.22z	H.221	H.222.1
						H.221
Signaling	H.230	H.245	H.230	H.230	H.230	H.245
	H.242		H.242	H.245	H.242	
Multipoint	H.243	N/A	H.243	N/A	H.243	N/A
Encryption	In draft revision	H.233	By reference to H.320	TBD	H.233	N/A
	H.233	Adapted in H.324			H.234	
	H.234	H.234				

H.320

Formally developed by the CCITT (now the ITU-T), H.320 is also known as the Px64 or H.261 standard.

H.320 is a Recommendation of the ITU-T for compression. It can be a video system's sole compression method or its supplementary algorithm, used instead of a proprietary algorithm when two dissimilar codecs need to interoperate. H.320 includes a number of individual recommendations for coding, framing, signaling, and establishing connections. It also includes three audio algorithms: G.721, G.722, and G.728.

A feature built into the H.320 standard is the ability to seamlessly translate between the NTSC video format (used in the United States and Japan) and the PAL video format (used in most of Europe). The standard covers data rates from 64 kbps to 1.544 Mbps, and up to 2.048 Mbps in Europe.

Most codecs adhere to this standard. When a videoconferencing unit is connecting to an "unlike system" (another vendor's), a standards-compliant mode is usually used. Proprietary modes still exist, and they may even offer more capabilities and improved video quality when used between systems from the same manufacturer.

H.324

Recommendation H.324 describes terminals for low-bit-rate multimedia communication, utilizing V.34 modems operating over the General Switched Telephone Network (GSTN). H.324 terminals may carry real-time voice, data, and video, or any combination, including videotelephony.

H.324 terminals may be integrated into personal computers or implemented in stand-alone devices such as videotelephones. Support for each media type (voice, data, and video) is optional; but if they are supported, the ability to use a specified common mode of operation is required, so that all terminals supporting that media type can work together. H.324 allows more than one channel of each type to be in use. Other Recommendations in the H.324 series include the H.223 multiplex, H.245 control, H.263 video codec, and G.723.1.1 audio codec.

H.324 makes use of the logical channel signaling procedures of Recommendation H.245, in which the content of each logical channel is described when the channel is opened. Procedures are provided for expression of receiver and transmitter capabilities, so transmissions are limited to what receivers can decode, and so that receivers may request a particular desired mode from transmitters. Since the procedures of H.245 are also planned for use by Recommendation H.310 for ATM networks, and Recommendation H.323 for non-guaranteed bandwidth LANs, communication connections with these systems should be straightforward.

H.324 terminals may be used in multipoint configurations through MCUs (Multipoint Control Units), and may work with H.320

terminals using ISDN service, as well as with terminals on wireless networks. MCUs and other nonterminal devices are not bound by the requirements in this Recommendation, but they should comply where practical. Connecting visual telephone systems using ISDN service (known as the H.320 series of Recommendations) and connecting mobile radio networks (known as the draft H.324/M series Recommendations) are also covered.

NOTE

An MCU is videoconferencing equipment that allows multiple individual videoconference units to connect together to form a multiparty videoconference session.

Figure 7.1 shows a generic H.324 multimedia videophone system. It consists of terminal equipment, a GSTN modem, a GSTN network, an MCU, and other system operation entities. H.324 implementations are not required to have each of these functional elements.

FIGURE 7.1:

An H.324 multimedia system

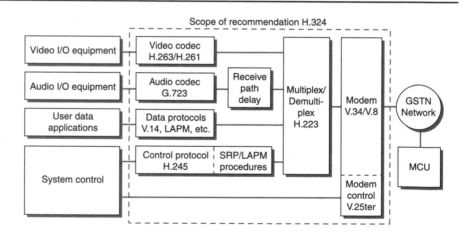

NOTE

The ITU Telecommunication Standardization Sector (ITU-T) is a permanent organ of the International Telecommunication Union. The group is responsible for studying technical, operating and tariff questions and issuing Recommendations on them with a view to standardizing telecommunications on a worldwide basis. The World Telecommunication Standardization Conference (WTSC), which meets every four years, establishes the topics for study by the ITU-T Study Groups which, in their turn, produce Recommendations on those topics. ITU-T Recommendation H.324 was prepared by the ITU-T Study Group 15 and was approved by the WTSC.

T.120

T.120 is a standard for audio-graphics exchange. H.320 provides a basic means of graphics transfer, but T.120 supports higher resolutions, pointing, and annotation. Users can share and manipulate information much as they would if they were in the same room. T.120 allows audio bridge manufacturers to add graphics to their products in support of a wide range of applications.

Video Quality

ISDN is required for desktop videoconferencing because the transmission speed is essential for acceptable video quality. Currently, picture quality depends on two factors: hardware quality and software functionality.

Video quality remains a thorny issue. To compete with television quality, a videoconferencing unit would require ten ISDN B channels, use video compression, and run on a processor with more power than the fastest Pentium Pro machines.

Compounding these issues is the audio channel. Even though the audio usually occupies its own channel on the ISDN line, compression can degrade the quality of the audio and make the video and audio quality a trade-off. The deficiencies in video quality can be partly offset by a good-quality audio connection. Again, compression rears its ugly head, sometimes creating a response lag, causing a loss of synchronization between audio and video data streams. The loss of synchronization may even create a slight echo.

Videoconferencing Communication Lines and Services

Between sites, many types of communication lines and services are available from the phone companies, including dedicated and switched connections. Switched digital lines, like ISDN and switched 56 kbps (kilobits—or thousands of bits—per second) lines are popular because they are cost-effective, efficient, and flexible. ISDN lines were developed to support all kinds of digital signals, including video, audio, and data, and handles them all equally well.

Within your organization, an interactive multimedia videoconferencing system can use your LANs or WANs to exchange data, but many are too slow (with too little bandwidth, or too small a pipe) to handle anything more than low-quality videoconferencing. More and more organizations are enhancing their networks with fiber-based networks and network backbones, dramatically increasing the capacity of their infrastructure. The FDDI (Fiber Distributed Data Interface) standard is popular for organizations that want to add bandwidth to their LANs and WANs, and it has plenty of capacity for medium-quality videoconferencing.

NOTE   Fiber Distributed Data Interface (FDDI) is a set of ANSI/ISO standards that define a high-speed (100 Mbps) LAN standard using fiber-optics as the transfer medium. FDDI allows a maximum length of about 60 to 120 miles and a total of 500 to 1000 stations. The two-core fiber provides two rings: a primary ring and a secondary ring for backup.

Videoconferencing most commonly uses 128 to 384 kbps, or one to three Basic Rate Interface ISDN lines. If available, multiple ISDN lines can be dialed to achieve higher data rates. The additional lines cost more, but they provide higher video quality. An inverse multiplexer (IMUX) is used to combine the bandwidth of the multiple lines, to allow the codec to operate at the higher data rates. Ascend, Teleos, and Promptus are all companies that make inverse multiplexers for videoconferencing applications, as well as for other data applications where low-cost, flexible, dial-up bandwidth is required. See Chapter 3 for more information about inverse multiplexers.

Although most videoconferencing occurs over terrestrial lines (those on the ground), satellite transmissions (or VSAT, for Very Small Aperture Terminal) are also used, particularly for distance-learning and business applications. Satellites offer the unique ability to reach many locations cost-effectively, and they are sometimes the only alternative in remote areas that are not covered by terrestrial lines.

Types of Videoconferencing Systems

Desktop systems are usually built on an IBM-compatible PC platform based on a 486 or faster microprocessor. The codec usually comprises an expansion board or boards, a camera, audio system, and Windows-based software. Connection to an ISDN line (or other type of digital line, such as switched 56 or fractional T1) is usually required to accommodate the video transmissions. During a call, you can see a moving image of the people on the other end, you can hear them, and perhaps most useful, you can share PC files and applications with them. The video quality on these systems is not as good as on larger, group systems, but it continues to improve.

Some of the desktop systems run up to only 128 kbps, and others run up to 384 kbps. As with the group systems, 384 kbps offers better video and audio quality, and faster data- and file-sharing capabilities.

NOTE   The new H.324 standard, described earlier, allows similar capabilities over a regular phone line, using a modem running at 28.8 kbps. However, the compression on these systems is so severe that the resulting video and audio quality is not usable for many applications.

Roll-about systems are complete videoconferencing packages contained in a wheeled cabinet. Designed for small- to medium-sized groups, roll-abouts are the most common type of system in use today. Usually, one or two monitors sit atop the cabinet, along with at least one camera mounted on a pan/tilt head, the audio system, the control system, and the codec.

The audio system consists of an echo canceler, microphones, speakers, and amplifiers. The control system provides the meeting participants with control over the video images, camera orientation, audio levels, and other peripherals. The camera in the roll-about captures the assembled meeting participants, and it can be remotely controlled to select varying views of the room. Convenient presets are also available to easily switch between commonly used views. And a graphics or document camera is commonly used to share documents, charts, maps, objects, and other graphics.

A dual-monitor system can typically display a combination of live video from the far end, a captured still image, and/or a computer application on each monitor. A single-monitor system uses picture-in-picture to show more than one type of display. While a still-image signal is being sent to the remote location, the motion video will momentarily freeze, until the transmission is complete.

Room and built-in systems include all the same equipment found in a roll-about, but instead of residing in a cabinet on wheels, they form a semipermanent or permanent installation. A built-in system can reside on shelves behind a facade wall, creating a permanent look that is preferred for some applications. Although the capabilities of these two types of systems are similar to roll-abouts, they often have more peripherals and are more customized to particular applications.

Applications for Videoconferencing

The following sections describe some of the applications for video-conferencing in two main areas: business and education.

Applications in Business and Industry

Videoconferencing is about making work easier and more productive by decreasing executive travel time and putting together people from different geographic locations almost instantly with just a micro-phone and a computer. As corporate LANs improve in speed, increase in number, and fall in cost, the corporate infrastructure is at a point where it is cost-effective to distribute video inside a company. In other words, videoconferencing can occur inside a corporation for the same reasons that people need to meet nationwide: cost and con-venience. ISDN offers the capability to extend this application out-side the corporate "walls" to the employee, contractor, or consultant.

Here are some statistics recently released in a research study of 663 corporations, *The Visual Communications Business Survey*, by the Pelorus Group (phone number: 908-707-1121). Companies respond-ing to the survey predict that within five years, 14.9 percent will have installed 100 or more videoconferencing units within their compa-nies, and another 83.4 percent will have some form of videoconfer-encing equipment installed. Within the next five years, companies offering their employees access to in-house videoconferencing will jump by 158 percent. Finally, companies that have more than

100 videoconferencing units installed at their sites will increase by 263 percent!

We've entered the era of "high-tech, high-touch" business. Today's competitive environment requires that corporations "listen" to their customers. Constant, fluent communications with customers and vendors are necessary for survival. Also, downsizing or "right-sizing" trends mean fewer heads to cover the same list of customers. Video-conferencing gives customers the "high-touch" or hand-holding they need without taking executives out the office (and out of touch). Staying close to home makes problem solving easier because specialists are within arm's reach. Problems get solved quicker.

Many times, high-tech, high-touch means getting closer to your customers. Companies are responding to their customers by going national, creating regional offices, distributed sales forces, on-site assistance, and the like. In Chapter 6, we examined another trend, telecommuting, which is contributing to the growth in videoconferencing. Letting employees work from home reduces overhead, helps retain talented employees, and improves employee's productivity and morale. Videoconferencing is the "glue" that helps dispersed workers stay in touch, share information, and collaborate on projects.

TIP

How much money can you save through videoconferencing? Every boss wants you to prove your new technology suggestion with an ROI (return on investment) analysis. Shayne Phillips of Comtech has created a handy tool on the World Wide Web. Fill in the blanks of his ROI Calculator and instantly see how much your company can save (or is saving) in a month or a year by reaching out and touching clients, customers, and branch locations. The address is: http://sa.comtech.com.au/shayne/vconf7.html.

Dataconferencing

Videoconferencing packages often include software that permits participants to share data as well as video and voice. Dataconferencing software is a tool that provides systems with the ability to share documents and the participants the ability to mark up "shared" documents—together.

> **NOTE** The dataconferencing standard is T.120, described earlier. This standard defines a series of communication and application protocols and services providing for real-time, multipoint data communications.

In some cases, users may find that video contact is not necessary and may share documents without video. This dataconferencing application is known as *whiteboards*. Whiteboards are interactive, stand-alone applications that allow two or more users to share files, such as a sales spreadsheet, presentation, or proposal. Together, each user may select a pointer and mark up the shared document, similar to an erasable pen on a white board. Sharing documents means that two or more participants can change figures on a spreadsheet, insert or delete paragraphs, even use an application that is not at their location or machine.

Applications in Education: Distance Learning

Distance learning refers to linking a teacher and students in several geographic locations via technology that allows for interaction between them. It was initially designed to extend educational resources into geographically remote or rural areas. However, interest has grown in using the technology to share scarce resources in urban areas as well as to meet the needs of students who cannot reach traditional classrooms. Researchers have concluded that distance-learning facilities overcome more than simply distance. According to a recent study, the top three reasons that students report enrolling in television-based distance-learning courses at the college level are time constraints, work responsibilities, and family responsibilities.

The prohibitive costs associated with many distance-learning programs have kept these benefits from reaching a wider public, however. Distance-learning programs are usually based on instructional programming delivered by satellite or through an Instructional Television Fixed Service (ITFS) network. Both methods require expensive equipment at the school and in the delivery system, and both allow for only limited interaction between teacher and student or among students.

Recent efforts to conduct distance learning over ISDN have been successful. According to a recent report, "ISDN offers students and teachers in distance-learning programs an interactivity level not available in ITFS or satellite programs without adding to the cost of the program." For these and other reasons, the researchers concluded: "ISDN has proved extremely efficient and effective technology for on-demand delivery of educational services to any region offering digitally switched phone service."

Videoconferencing and Video Telephony Case Studies

Video telephony is both an application in itself and a component of many other innovative uses of multimedia telecommunications. Due to great advances in signal processing, even analog phone lines can be used today to send and receive slow-speed video images. ISDN-based video telephony and videoconferencing are now available and affordable.

When Barbara Bush videoconferenced from the White House with children at a Baltimore hospital at Christmas, she was using an ISDN connection. When a group of Lawrence Livermore Laboratory scientists work at home, ISDN enables them to use their personal computers, without a modem, to tap into the lab network and get a data connection 27 times faster than normal. In the following sections, you will read about many other examples of real-life videoconferencing applications.

Each of the video-based applications discussed here runs over Basic Rate Interface (BRI) ISDN, the standard 2B+D configuration described earlier in the book. In some of these applications, a video connection at 64 kbps or 128 kbps is linked to a wider bandwidth trunk (a Primary Rate Interface, or PRI connection), yet the transport to the end users remains a BRI connection. These types of applications are of interest to those considering the functionality of single-line ISDN. The case studies presented here demonstrate how the availability of BRI connections can enhance the reach and flexibility of private networks and allow traveling employees, contractors, and affiliates to connect to a corporate network from beyond the normal reach of the network.

Education Case Studies

The following examples suggest that ISDN is a viable, effective technology for distance learning, with shared workscreens, videoconferencing, and access to off-site resources in many settings.

NOTE

ISDN could also enhance the future availability and value of educational resources. Congress has passed legislation calling for the creation of a National Research and Education Network (NREN), designed to link educational institutions, government, and industry in every state. Among its purposes, Congress sought "to promote the inclusion of high-performance computing into educational institutions at all levels." The investment needed to actually connect every one of the nation's 84,500 public schools and 24,000 private schools is far beyond the resources available in the NREN legislation. If ISDN were widely available, it could substantially leverage the value of the government's investment by enabling schools, especially at the K-12 level, to attach to the NREN and reap the benefits of this high-performance network.

Distance Learning and Videoconferencing in California

California State University, Chico, in partnership with AT&T and Pacific Bell, has completed two successful trials of ISDN as a delivery system for distance learning. In May 1992, ISDN was used to link fifth-grade classes in three elementary schools in the Chico area. During the trial, students shared slide show presentations and participated in conference calls that included video, data, and image transmissions. Through this system, students and teachers were able to engage in real-time, interactive, two-way audio and video communication, view a laser disc video-clip display, and send and annotate graphic images or text.

In March 1996, scientists at the San Juan Institute in San Juan Capistrano gave a free public lecture on the newly discovered Comet Hyakutake. The event was connected via videolink to three locations elsewhere in the state, including Bryant Elementary School in San Francisco.

Pacific Bell Education First hosted a videoconference debate in Anaheim in April 1996. About 20 Mission Viejo High School juniors, who are members of the International Baccalaureate program, debated with their peers from Winston Churchill Secondary School in Vancouver, British Columbia. The Canadian students were at their facility called Science World, which is home to many internationally recognized Pulitzer prize winning scientists. The debate topic was: "Fertilized Eggs — Is It Wrong to Sell Them?" A lively discussion ensued between the participants in the two locations, who were communicating via videoconferencing using three ISDN lines to interact. When questioned after the session, the students were enthusiastic in support of having the technology to talk with their colleagues in another country and were disappointed that not all their views were able to be shared (due to time constraints)!

The new musical, The Party's Over, was presented on April 27, 1996, in the Music Building of Los Angeles Harbor College. The show, a first-of-its-kind event, also was transmitted via videocast to a group of adult education students at Northeast Illinois University.

Library Legislation Day, Tuesday, April 30, 1996, was a day when school librarians, public librarians, and library supporters visited Sacramento to discuss key issues with their legislators. Two Education First technology demonstration sites, Martin Luther King Middle School in Seaside and Pasadena Public Library, had videoconferences to the capitol's videoconferencing room. The Martin Luther King Middle School connection was most impressive because they had about 20 to 25 people at that end. At the Sacramento location, County Public Librarian Dallas Shaffer, the President of Monterey's Library Foundation, and several others sat on either side of Assemblyman Bruce McPherson. Library supporters from both locations asked Mr. McPherson many questions.

NOTE

One favorite question during the Library Legislation Day videoconference was from a fifth grader who simply wanted to point out the importance of having his school library open longer, since students already were at school and often their parents could not take them to their public library. He clearly described how many parents were without transportation or worked odd hours and couldn't easily get their children to the library—so strengthen the school library! The young boy's comments made a real impact on the Assemblyman—as did videoconferencing from Seaside.

Irvine's Vista Verde Elementary School students visited California's State Capitol in Sacramento in May 1996. Students in Irvine interacted with Sacramento via a videoconferencing system donated by Pacific Bell's Education First initiative for this event.

Also in May 1996, students from Bryant Elementary School in San Francisco participated in South Korea's "Kid Net Day" via videoconference. "We are all really excited about this international event. Our students studied and prepared for this first-ever opportunity to meet and talk live with their fellow students in Seoul," said Virginia Davis, the Technology Resource Teacher at Bryant.

Two Distance-Learning Networks in North Carolina

Appalachian State University, AT&T Network Systems, and Southern Bell have built an ISDN-based distance-learning network that delivers interactive voice, data, and video to three North Carolina schools. The ten-year project, called "Impact North Carolina: 21st Century Education," was touted as "one of the first in the nation to deliver interactive video instruction through existing copper phone lines." The system transmits interactive voice, data, and video at 112 kbps to two elementary schools and one high school in Watauga County. The Impact North Carolina system will give K-12 students access to remote lecturers, university libraries, and other distant resources. The system will also be used to improve teacher training, student teacher supervision, and continuing education at Reich College of Education, a major regional center for educating teachers.

In the Research Triangle Park area of North Carolina, the North Carolina State University Center for Communications and Signal Processing, BellSouth, Southern Bell, GTE, IBM, Northern Telecom, and the Wake County Public School System are developing SCHOOL-NET. This is defined as "a project to demonstrate the enhancement of public education through advanced telecommunications technologies, specifically ISDN." Among the functions that SCHOOLNET plans to provide are video learning and distant instruction; electronic access to library materials; faculty support for exchange of curriculum materials and teaching aids; and administrative support for scheduling and staffing purposes.

Project Homeroom in Chicago

Project Homeroom, an initiative "designed to improve student thinking, learning and computing skills," is a partnership among six Chicago area schools, Ameritech, IBM, Illinois Bell, Prodigy, AT&T Network Systems, Central Telephone Company, and Eicon Technology Corporation. More than 550 students are participating in the project, which uses PCs, multimedia software (with CD-ROM, video, voice, and text features), and online services over phone lines supplied by Illinois Bell.

Students access online homework correction, instruction, and tutoring, as well as computer communication with teachers, among other features. Seventy-six of the participating students from Stagg High School in Palos Hills, Illinois, use the system over ISDN. ISDN allows these students to exchange text, pictures, and calculations up to eight times faster than other students.

Learning Technology in Nashville

In Nashville, Tennessee, students at Carter Lawrence Middle School and Meigs Magnet Middle School can work together and with the Learning Technology Center at Vanderbilt University over ISDN. The pilot project, a joint effort of South Central Bell, Northern Telecom, Vanderbilt, and the Tennessee Public Service Commission, uses voice, video, and screen sharing technologies to enable students to see and talk with other students or with faculty at Vanderbilt. They can also share documents, graphics, and other information.

Government and Politics Case Studies

Videoconferencing has its place in government and politics, too. The following are some examples of how this application is used by government agencies and political groups.

California Scientists and Engineers

The Jet Propulsion Laboratory in Pasadena, California, and the Lawrence Livermore National Laboratory in Livermore, California, are each using ISDN for desktop videoconferencing to conduct business among scientists and engineers between the two sites. The system also allows screen sharing and file transfer capabilities during the videoconference. Project managers at Jet Propulsion Laboratory have also been testing the desktop multimedia system independently.

NOTE

Stan Kluz, an ISDN expert at Lawrence Livermore, recently hooked the first group of ISDN users off site into the laboratory's computer network. Kluz said that through this arrangement, 12 scientists who live near the University of California at Berkeley can use their computers at home and have access to data at 64 kbps. With speeds that fast, the scientists can manipulate huge amounts of data and see their problems displayed in three-dimensional graphics on their home computers. Kluz sees the future of telecommunications, and it is ISDN. He says that videoconferencing on all ISDN-equipped computers at Lawrence Livermore will be available soon. With nationwide interconnection agreements, he hopes to see distance learning in which a class in, say, nuclear physics at the Massachusetts Institute of Technology could be hooked to the computer of a Lawrence Livermore scientist, who can take part in the class.

Videoconferencing in the Navy

The Naval Air Warfare Center Weapons Division at China Lake, California, the Navy's largest research and development project, is using Basic Rate ISDN for desktop video teleconferencing at the facility, where offices can be as far as 20 miles apart. They plan to extend the ISDN-based system to create video links to AT&T Labs in New Jersey, and the National Institute of Standards and Technology in Maryland.

Citizens of California Videoconference

A collaborative effort among the League of Women Voters, the California Constitution Revision Commission, and several schools and libraries are allowing Citizens of California to share data and information regarding hot political topics. A six-week series of interactive videoconferences was conducted late last year to gather public reaction to preliminary recommendations of the California Constitution Revision Commission. Citizens meeting in Mendocino, Fresno, Saratoga, Santa Ana, and Pasadena were able to see, hear, and interact with commission members located in Sacramento.

Citizens, Government, and Videoconferencing

The California Constitution Revision Commission was established in 1993 to recommend ways to improve responsiveness and efficiency of state and local government, increase flexibility, and enhance fiscal integrity. Input from the meeting will be part of the final recommendations of the Commission to the legislature for inclusion on the November 1996 ballot.

"We are excited to be pioneering the use of videoconferencing to establish a dialogue between citizens and government on this issue of vital importance to the state's future," said Gail Dryden, state director of the League of Women Voters. "We believe this technology holds significant promise for facilitating greater citizen involvement with government, particularly for rural and suburban communities where public hearings on state and national issues are less frequently scheduled."

"Schools, libraries and community colleges have traditionally functioned as centers of their communities, serving as sites for town meetings and other civic activities, " said Dan Theobald, community applications manager for the Education First Initiative. "Now these institutions are expanding into a new role as telecommunity centers, helping local citizens connect to the information superhighway by offering public access to information technology when it's not being used for student instruction," Theobald said. "Beyond video democracy, school-based technology could support community applications in the areas of health, arts and humanities, economic development, and social services."

Each video "town meeting" was held in a school, library, or community college, which had been equipped with videoconferencing equipment and ISDN lines provided by Pacific Bell's Education First program. This program is a Pacific Bell initiative to help connect

California schools and libraries to electronic resources and communications by the year 2000. This series of video meetings was part of a larger effort by the Commission and the League to conduct forums in communities where in-person meetings could not be scheduled, thereby increasing citizen involvement throughout the state.

Industry Case Studies

And of course, videoconferencing has been used in industrial applications. Here are just a few examples.

- Hitachi America, Ltd. has installed Basic Rate ISDN at many of its North American locations to enable videoconferencing between Hitachi locations in the United States and for international meetings with executives in Japan. People involved in the videoconferences can simultaneously exchange data files and fax messages over the same lines.

- In France, France Telecom closed down its 2 Mbps videoconferencing network service and replaced it with dial-up equipment over ISDN service. Subscribers now dial national and international calls using the French "Numeris" network at 128,000 bps and will be able to participate in multipoint calls with up to 16 locations.

- At General Motors' Troy, Michigan facilities, the car manufacturer is testing ISDN links to the company's private network, the world's largest. A Sun workstation running CAD/CAM design software has been outfitted with a video camera. The configuration allows users to share screens of CAD/CAM and whiteboard displays over one B channel and run desktop videoconferencing on the other B channel.

Can You Afford Desktop Video?

Although some desktop video solutions were available earlier, Intel's introduction of its Proshare product in early 1993 spurred the interest in the desktop video market. The original Proshare product allowed data collaboration and videoconferencing on the same PC and had a price that many companies could afford. The introduction of this product challenged the other companies to speed their desktop video solutions to market.

Since that time, there has been pressure to provide a desktop video solution for under $2,000. In the current market, some desktop video products are listed for over $4,000. However, it will not be long before you will see a $500 video multimedia option for most PCs. See Chapter 12 for information about Proshare, as well as PictureTel's Live200 series, the CLI's new desktop system, and other videoconferencing systems.

CHAPTER
EIGHT

8

About Remote LAN Access (RLA)

- ■ A definition of RLA

- ■ Advantages of using RLA applications

- ■ The conversion of the home office for ISDN services

- ■ Corporate remote-access policy

- ■ RLA and security issues

- ■ Some RLA case studies

- ■ RLA tariff considerations

In Chapter 6, we talked about the changing corporate work place and the growth in the number of workers who are telecommuting. Remote LAN access (RLA) is an extension of this telecommuting trend.

RLA—being able to log in to your network from a remote location, such as your home office—is another application area for ISDN connections. RLA is a large subset of the fast-growing internetworking industry. As more and more businesses install LANs and interconnect them with other LANs, the need for workers at remote locations to access LANs increases. Providing remote access to the LAN is the next step for many growing companies and their respective networks.

How Is RLA Different from Telecommuting?

RLA's key difference from telecommuting is that the communication includes a LAN connection. Because a LAN is involved, there are more design considerations to address for RLA applications. For example, you need to consider security, equipment standardization, and the setup of the home office workstation. We'll discuss these design issues in more detail later in the chapter.

Today's corporate and mobile workers can fit many different profiles. RLA users can range from people who occasionally work from home to full-time, contract programmers who download programs and Web server home pages to the headquarters office on a daily basis. Analysts and market research groups have struggled to classify these people who "work at home," and in some cases, have oversimplified these classifications. The MOHO, ROHO, and SOHO terms, for the mobile, remote, and small office home office workers, apply to the RLA users as well as other telecommuting groups. However, once you connect to a LAN system, there are specific elements of your application and its use that you need to consider.

With RLA, there usually isn't exclusive use of just one application. In fact, the trend is toward using multiple applications, such as videoconferencing, faxing, and online service access, along with RLA applications. This makes remote LAN access more complex to configure and set up than telecommuting applications that don't involve LAN connections. Figure 8.1 illustrates the difference between a typical RLA and telecommuting setup.

FIGURE 8.1:

Telecommuting versus RLA applications

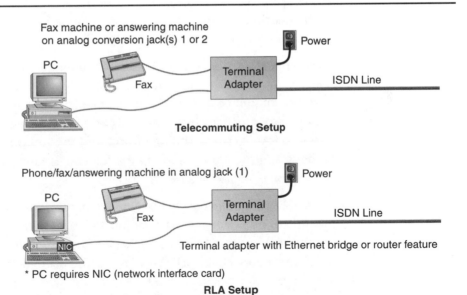

ISDN as the LAN

An interesting bit of ISDN history has to do with the evolution of ISDN and LAN systems. Some RBOCs did some development work and marketing that would have you believe that ISDN could replace your LAN system. The idea was that ISDN could use the UTP (unshielded twisted-pair) cable as the transmission path to your end-user stations. Using ISDN would allow companies to create larger LAN systems.

As you probably know, this application never really took off. It was only viable if you could find affordable ISDN terminal equipment and also the ISDN service available to connect to your branch offices.

In addition, Ethernet at 10 megabits per second (Mbps) was becoming a popular standard, and UTP cable was able to transmit the 10 Mbps to the desktop. Many corporate LAN users opted for the Ethernet solution in the building or campus, and they did not need to run everything into the phone company's central office.

Most companies found that simple LAN hubs and Ethernet-based systems were more available and easier to implement. Also, the ISDN solution did not immediately solve the congestion problems or provide the hub-control functions that most LAN administers needed to manage a growing LAN system. The bandwidth of Ethernet became a standard benchmark; most users wanted the full use of Ethernet and not just a smaller pipe version like ISDN.

Now, ISDN's features allow for dynamic bandwidth allocation and connection of two distant LAN systems. ISDN is being used instead of a leased line for two LAN systems that need connectivity but do not need a full-time connection. ISDN services can be used for full-time operation and connection between two LANs, or simply as backup or additional bandwidth to an existing private leased-line connection. This setup is popular when the LANs are close together and the usage costs of ISDN aren't more than leasing private lines. The quick ISDN connections and disconnections allow for most LAN systems to operate without user-perceived interruptions.

Why Have Remote Access to a LAN?

Whether you're a contractor who needs access to your clients' LANs or an MIS manager of a company that has a LAN (or a WAN) and off-site employees who need access to it, RLA is for you. The following sections summarize the benefits for corporations, employees, and customers.

Corporations and RLA

For corporations, RLA can have benefits such as these:

- Distributed LAN systems allow for improved intracompany communication and productivity.

- Remote-access capabilities increase the company's flexibility in recruiting employees, temporary consultants, and other workers.

- Using RLA lowers costs for providing network access and maintaining office space for employees.

- RLA can be a low-cost solution for small branch offices that cannot afford the cost of installing a LAN system.

- RLA use requires adherence to a standard LAN system policy, and this standardization reduces LAN support and equipment costs.

- Addressing the security issues linked to allowing remote access to your system improves your company's overall security policy.

Remote Employees

The benefits of RLA applications for employees are fairly obvious:

- RLA applications make corporate resources and information more readily accessible. Critical data and databases on corporate servers are available to you 24 hours a day.

- Remote access gives you flexible work hours.

Customers and RLA

Customers like the use of RLA applications for several reasons:

- LAN access allows for customer contact personnel to improve their responsiveness and service level.

- With RLA, it's much faster and easier to find customer files or information when the LAN is required for data retrieval.

RLA Buzzwords Brief

Many of the terms that have been defined in the previous chapters in this part also apply to RLA applications. The following sections define some other common terminology.

Client/Server Computing

In discussions of remote LAN applications, you will hear a lot about the *client* or *terminal* and the *central host* or *server*. When referring to a remote LAN access environment, the client is the remote PC that is calling into the central host or server on site. This is the basic architecture for the connection to a LAN in a corporate network or business environment.

When talking about RLA programs or network configurations, you will hear the terms *node* and *hub* used interchangeably. The host, node, or hub location is the place, computer, or server where all the remote users call to access the LAN system and get the information or files that are needed.

Client/server computing is a term used by internetworking companies and software system developers. As shown in Figure 8.2, the client server architecture refers to the design of a computer system that allows many users (clients) to access the same information or central computer (server). This design requires that the client terminal, which in most cases is a PC, and the server computer (usually a more powerful PC or mini-computer) share the task of processing information. The client and server may share the application software and the database information. There are many variations, but in general, client/server designs center on the wide-area network (WAN) that connects the client and server locations.

FIGURE 8.2:

A client/server architecture lets many users (clients) access a central computer (server).

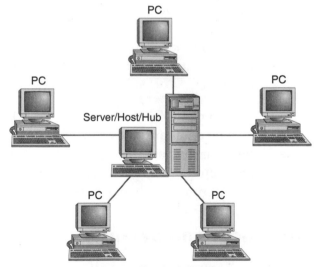

In the past, most of the systems that allowed remote access were based on the use of a leased private line only. This was usually very expensive, and the charges for the more distant locations were so high that the companies did not allow them to connect to the corporate computer. The lower cost of computers, more intelligent software, and the more options for network connections have made the client/server architecture more feasible and attractive to businesses.

For example, the way that a bank branch uses the bank's database of information is an example of a client/server application. If your account is based at the branch you visit, the teller can use a PC terminal to look up your information for your account and give you a statement immediately. If your account is with another branch, the teller's PC would format the request and send it to the bank headquarters computer center. The network would provide the connection between the teller PC and headquarters computer center.

In another type of computer system, the teller's terminal (not a PC) could have a direct connection to the large mainframe computer in the headquarters center. The request would go the mainframe, and the terminal would do very little processing. The ability to put some processing in the teller's PC provides better customer service and reduces the amount of time that the teller needs to connect to the headquarters computer.

What client/server means to the MIS people of the world is connecting company sites and employees to the disbursed remote and headquarters sites in a client/server distributed computing company environment. This also creates the need to connect more sites of a single company than ever before.

Intranets and Security

Intranet is a term that refers to the network of networks that many corporate networks are becoming, with connections to the Internet as well as multiple corporate department networks.

The LAN system is a local network. When companies connect their different LAN systems at different sites, that creates a network of networks, or intranet, for that one company.

When companies connect their networks to the networks of other companies and to the public Internet, the term intranet applies on a grander scale. This type of intranet, with connections outside a particular company, raises special security concerns. The companies must make sure that there is not any unauthorized access and use of their network.

As an example, Figure 8.3 shows an intranet of a company's corporate sites, a vendor, and the public Internet. In this setup, you would want to make sure that your employees had access to public Internet resources, but you wouldn't want to allow access to the company's confidential financial or customer data from the Internet or vendor's network.

FIGURE 8.3:

An intranet connects the networks of different companies, as well as the public Internet.

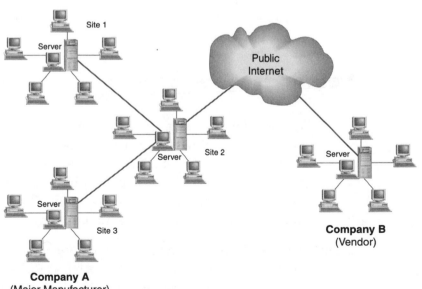

This security and the measures to provide it are what *firewalling* your intranet is all about. The *firewall* is the security access or protection that will let only authorized users onto your network or intranet. Companies can set up separate groups of users with different security levels. For example, the human resources group might have access to payroll and personnel records but not to the billing records or sales information. Passwords and other security procedures allow network administrators to control access to company information.

The *dial-back* method of security is one fairly economic way for a small implementation of RLA. The user calls into the host or central site and requests a connection. Then the user hangs up, and the central site computer calls the user back at the user's home or office, using a preprogrammed phone number. This is just one simple level of security.

Security is very important in an RLA program. Company information is very sensitive and must be protected, just like equipment and company business plans. You will hear more about firewalling as more people use the Internet for transactions and business communications.

Networking Protocols

The standard protocols that an RLA program needs to support are Ethernet, IPX (Novell), Appletalk (Macintosh), or TCP/IP. These protocols must run over ISDN and interface with your PC and the central host location. This protocol support is the main factor that makes RLA different from basic telecommuting. It also creates the need for more sophisticated software and hardware in your terminal adapter.

Ethernet is one of the most popular standards in the LAN industry. It has been extended to switched Ethernet and Fast Ethernet (100 Mbps), so it will continue to be a viable networking option.

An NIC (network interface card) is a hardware card necessary to interface your computer to a network. Usually, the term refers to an Ethernet interface card that a PC needs to connect to an Ethernet LAN system. In the RLA environment, the PC in your house will need an Ethernet card to connect to an Ethernet LAN system at the host or central site.

Routers and Routing

In order to send information from one network to another, that information must be *routed* along the proper path. This routing is provided by the *router*, a computer hardware device that may be an internal or a stand-alone unit. Basically, the function of a router is to provide a path from a node on one network to a node on another network.

When the router connects WANs, it must find paths over widely distributed networks. A WAN router needs to support protocols suitable for long-distance access. To support a network connection, your terminal adapter will need to perform routing functions as well as basic ISDN protocol functions.

RLA Design Considerations

The use of RLA applications increases the number of physical and logical connections to your network. The remote LAN user needs to appear to your LAN node or hub like another local LAN user. All the elements of your system must conform to LAN protocols and interconnection standards.

For your RLA setup, you need to consider several design issues:

- How does the home office need to be changed?
- Is there a corporate or company policy regarding remote access to your network?
- How are LAN protocols handled?
- How can you convert RLA applications to use ISDN service?
- What security is needed to protect your network?

Today's Remote Access with Analog Telephone Services

Currently, LAN connections are made through analog service, at speeds in the range of 14.4 kbps to 28.8 kbps. This is barely acceptable for most users who are connected to a local LAN. The files, presentations, and graphic images for doing any business are getting larger and larger.

Many users complain of the time it takes to log on and get access to their corporate networks and then download files and graphics. This has led to the demand for corporate MIS staff and small business owners to provide some faster network connection for data transfer.

In the remote LAN access arena, this means moving to another network option that is faster than regular analog dial service. The next step for these users is a networking system that provides the switched dial connection and any-to-any connectivity. ISDN can fulfill these requirements. Connecting through ISDN service not only improves the functionality of the remote LAN access application, but it also provides the ability to make connections using the same North American numbering plan as analog phone service and allows you to connect to the same small sites as you do with the current network.

ISDN service will improve the data-transfer speed and the call connection and disconnection time. This will increase your productivity and decrease usage times for sharing information or files.

Tomorrow's Remote Access with ISDN Telephone Service

With the addition of ISDN service, the office communication options are even more complex than before. You will need to purchase a new piece of equipment for your office, called an ISDN terminal adapter (TA). A TA is a modem-like box that can support up to 128 kbps of data transfer (one B channel is roughly four times the rate of the fastest modem available). This implementation solves the slow communications rate problems, but leaves the phone, fax, and modem connected to one or more analog phone lines. You may ask, "Why do I still need a modem when I have ISDN service for data calls?"

Unless you have a modem, your ISDN-connected home office won't be able to communicate with many established modem sites, because ISDN data calls are incompatible with analog modem data calls. Normal ISDN terminal adapter equipment has no provision for simulating a modem and cannot exchange digital data using a voice-call connection. Therefore, you need your analog modem setup to maintain compatibility with the currently installed base of modems.

Having two different solutions for data transmission means that incoming data calls must arrive at the correct phone number for each type of call—either the ISDN number for ISDN data calls or the analog number for modem data calls. It's difficult to match the incoming calls to the correct phone number.

Currently, you need to configure your home equipment to take advantage of ISDN when you can. You also need to have a separate analog line for your analog equipment or connect your equipment to analog conversion ports on your ISDN terminal equipment.

Fortunately, most terminal adapter manufacturers have realized how convenient and important it is to still use some analog devices with ISDN. They have added at least one analog conversion port for this purpose. You should be able to use equipment such as a fax machine or phone answering machine for your office. This conversion port will usually support one phone extension per port.

> **WARNING** Some analog conversion ports do not support all the standard characteristics of analog phone lines. Before you make a purchase, make sure that the conversion port will support the analog equipment that you want to use at home.

In company or corporate environments, the standard setup is for the ISDN users to have one group of numbers and the analog phone users to have another group of modem pool numbers. Many of the hub equipment providers (Ascend, US Robotics, 3Com, and others) now offer equipment for the company central site that can handle both ISDN and analog phone connections. This will make it easier and more cost-efficient for everyone; carriers and service providers can use this equipment in their call-in facilities for many different applications or online access locations.

Eventually, you will not need to know what the other end of the connection can manage (ISDN or analog phone line), but you will be able tell that the online service is using faster communication and data-transfer speeds. This will also come about when the terminal adapter vendors integrate a modem function as a backup to your ISDN terminal adapter. We'll discuss this trend in Chapter 15.

You should be aware that there is some difference in the manufacture and price of terminal adapters when LAN protocol support and routing are provided. It is for this reason that the Ascend and Cisco equipment has been more expensive than the basic terminal adapters available from other companies, such as the earlier versions offered by Combinet (now owned by Cisco) and Adtran.

Fortunately, much like router prices, the cost for router functionality in terminal adapters will fall in the next year. The router market has seen prices decline by 50 to 100 percent in the last year with increased functionality in each router product. This cost drop will extend to ISDN terminal adapters with routing functions for RLA applications.

Is There a Corporate Policy for Remote Access?

The company's policies concerning remote access to your LAN will determine how your RLA applications are set up. Some companies simply do not allow any employees to access the corporate LAN from external locations. This ends the policy and technical issues fairly quickly. The company has decided not to sponsor remote employees or allow contractors remote access to corporate information or file sharing.

On the other hand, many contractors and programmers require a certain level of network access as a condition of taking the project or working for the company. Some consultants demand quick and responsive access to their client's information and management. In some cases, it seems that contractors and consultants receive better access and equipment than the employees. Many programmers have high-speed access and Ethernet connections directly to their company's or client's corporate site for quick exchange of software programs or code.

The remote or mobile worker may also be the sales executive or advertising agency who works part time on a specific project, or during off hours at home, and needs certain access at various times.

The special time and access requirements also create the need for support and help desk functions, which should be addressed in your remote LAN access policy. Remote users may be looking for support at 11:00 P.M. to turn in a project or proposal early the next day.

Problems will arise if users can't get access or download information when they need to.

The important point is to make sure that you gather and document your company policy for employees, contractors, and others to assess the demand for remote access to your LAN.

Your company policy also helps you decide on the makeup and size of your remote user group. You will need to assess the scale of your RLA program, who will be included, and the related requirements of these users. Since an integrated NIC isn't built into every PC yet (although one will probably be within the next two to three years), you will need to identify the equipment and standards for remote connections to the LAN.

Your policy should include profiles of your "typical" users and templates of the equipment configuration and ISDN service configuration. You'll also need to provide support for those employees at home, perhaps on a seven-day, twenty-four hour basis. See the section about corporate standards for telecommuting in Chapter 6 for more information about what your company's standards should cover.

NOTE Today, most phone companies offer extended ISDN coverage (7 A.M. to 7 P.M., with some weekend coverage) and most will get to the full coverage (7 days a week, 24 hours a day) as the deployment of ISDN volume grows.

What Are the LAN Protocol Issues?

The common LAN protocols, or corporate LAN interconnection methods, usually involve network protocol standards, such as TCP/IP, Novell's IPX, Ethernet, Appletalk, and PPP. You will need to know what your LAN system architecture will allow for your remote connections. See Chapter 3 for more information about network protocols.

The cost of the equipment for RLA will be more than for standard ISDN terminal adapters without the software to manage these protocols or LAN connections. The hardware and software for the RLA user will generally require more intelligent programs or routing capability. However, as mentioned earlier, the costs for ISDN terminal adapters with routing capabilities are declining.

The LAN functionality also increases the complexity of the ISDN connection to some degree. RLA applications demand complete compatibility of the systems at both ends of an ISDN connection. Users need to understand that the extra software features that you are getting for this application are the reason for the higher price of the terminal adapter and client software. The good news is that once the remote user is set up and has everything installed, he or she will never choose to go back to the analog phone line. Again, the call setup time and increased speed of data transfer will win over these remote LAN users.

Converting RLA Applications to ISDN

Most existing RLA solutions today are based on analog phone service, and many vendors have mastered the network connection issues using this standard dial service. Now companies are converting their systems to use ISDN service for their RLA applications. For example, Shiva Corporation, a hardware vendor who specializes in meeting the needs of the RLA market, offers equipment and related software for ISDN service.

The conversion of the application to ISDN service instead of analog phone service is not as simple as you might think. It requires changes in the client and server hardware, as well as changes to the software being used.

First, you must have ISDN-compatible hardware and software installed at the server site. Usually, you will need to replace your existing analog phone equipment, upgrade related hardware, and add phone lines, so that your system will be capable of ISDN connections. This equipment will be in addition to your network's current analog modem pool arrangement (as explained earlier, you'll want to keep your analog phone service and modems for some connections and for backup). You may also need to invest in some new security or gateway software that is compatible with your new ISDN architecture, network, and equipment.

Security Considerations

Security is an issue that is part of network access in general. Remote access adds another dimension to your security measures. In the LAN environment, you'll want to protect critical business and financial data with a firewall. This is becoming increasingly important as the corporate networks grow into larger intranets and people have access through public Internet connections.

As people become more knowledgeable about computer systems (consider today's college students, who are becoming computer science savants), there is more critical need for security measures and programs. These concerns also apply to analog modem access, but they merit special attention with ISDN because some of the existing security measures are based on the analog technology and interface and are not easily transferred to the digital ISDN world.

This situation is improving as the software and equipment vendors see their customers and markets moving to digital network options. Many developers are either creating new security measures or converting existing ones to ISDN network access. Soon you will see new ways to increase network security using methods that are more efficient for the user.

Since ISDN has the ability to forward the calling and called number, this implies that you can identify the location of the caller. This provides one layer of security for users who work from a single location. However, it does not work for the mobile user unless that user knows every site that he or she will work from.

> **NOTE** Previously, the ISDN interface itself was a security measure because you could not access the central hub or site unless it was an ISDN interface or equipment user for the connection to take place. This will not work in the future, as more and more users have ISDN service.

There are many layers of security, and your online service provider or company administrator needs to develop, design, and implement the levels required. The security designer must balance the requirements for security measures and ease of access for users. If users must go through a ten-minute routine to get to information or a service, they will not be eager to access that location or service.

We'll discuss security measures for RLA applications in more detail in Chapter 11. Just keep in mind that whenever you are dealing with remote-access methods, security will always be an issue. As the demand for remote access continues to grow, the industry will develop better solutions in software and hardware for ISDN connections.

RLA Case Studies

The need for RLA is an extension of the trend to move LAN users outside the traditional campus environment. The following are some examples of companies that have invested in RLA programs using ISDN as the networking option.

Hewlett-Packard Goes Mobile

For many years, Hewlett-Packard ran its work-at-home program using analog phone service. But in 1994, the workers began to demand faster and higher bandwidth service. The managers decided to try using ISDN services. They created a corporate program and infrastructure for the entire company to access their corporate backbone network and resources from many sites.

Hewlett-Packard is a worldwide company, and it needs worldwide connectivity. Although ISDN is not available everywhere, it will be used wherever it is available, both within and outside the United States. The LAN connection allows the employees to "plug" into the corporate resources and keep in contact with support groups and other employees.

The expansion of the users or clients into homes brings on new requirements for the Hewlett-Packard MIS groups. It is similar to the transition from using an office desktop PC to running a portable PC, with the related service and equipment issues. However, it appears that Hewlett-Packard's RLA program has increased the productivity of the workers involved.

Intel Plugs into RLA

Intel introduced the Proshare product in April 1994. This was a hardware (PC cards) and software package that allowed a PC to communicate with another PC. This communication package included videoconferencing and data collaboration capabilities, but it was not designed for LAN connections or Internet access.

As part of the Proshare product line, Intel introduced an RLA package shortly after it put the Proshare units on the market. The plan was to add this to the product portfolio of Proshare, and have the RLA product become as popular as the videoconferencing and data-conferencing packages.

Intel introduced another version of the Proshare product line that included local and remote LAN connections. The remote LAN package was the version that used ISDN as the network communication. Intel was the first manufacturer to involve the RBOCs and some long-distance carriers in the development of a different channel for the company's ISDN products.

The local LAN version was not accepted by most LAN corporate administrators because videoconferencing was not an important feature for most LAN users, and because it did not have a robust LAN management tool to balance the LAN bandwidth for the corporate site. Videoconferencing on most LAN systems will create havoc with other LAN users; without controls, the video application will take up the bandwidth of the LAN and cause congestion for other users.

Also, the original Intel LAN package focused on only the client or remote PC end. It did not have any support for the central host site equipment. You needed one-to-one equipment for each remote user, and users couldn't share the resources in a modem pool environment. Given the demand concern and this host condition, this package wasn't very popular in large corporate environments.

Intel has since downplayed the video features on the LAN and put more emphasis on the RLA capability, with ISDN on the client/ terminal end and also at the host site. There are now standard interfaces that can be used with other hub or router manufacturers for this configuration. Using standard LAN protocols has increased the options for the corporate MIS staff to use many different hardware options and avoid the one-for-one equipment requirements. This has also positioned Intel back at the desktop (versus hub or router environments), which is the company's strength.

Sun Microsystems' RLA Program for Programmers

Sun Microsystems, like many of the other members of the high-tech community, needs to provide remote access to corporate resources. However, the company also hires many developers and programmers on contract. These are temporary employees who work at home and at remote sites and who need some connection to Sun's headquarters site.

Typically, these contractors and programmers need to transfer large files rather than just text documents or e-mail messages. They send entire projects or software programs. Understandably, they have been frustrated with the analog dial-up option. Now some of these remote workers have been able to use ISDN service for their connections, and they appreciate the fast and reliable communications.

Silicon Graphics Gets Extra Bandwidth

Silicon Graphics is another company that employs many contractors and programmers that must transfer large files. In this case, the workers exchange a large number of graphics files. Some of the contractors and programmers have been able to send and receive these files using ISDN connections, and they have seen a great improvement in throughput.

ISDN is just one of the options for the Silicon Graphics employees. Other high-speed connections and some dedicated network services (such as full T1) are also available in some locations. The need to transmit graphics and animation, such as the images included in World Wide Web pages, is part of the communications trend that is creating the need for more and more bandwidth.

RLA Tariff Considerations

When you're using ISDN services for RLA applications, one of the main issues for the tariffs is the availability of the service to the sites or homes that need it. After you verify availability in your area, you need to decide how many people in your company or community are going to use ISDN for access to your LAN systems. Local versus long-distance usage is a major consideration, as it is with any usage-based network service.

Review the available packages of ISDN service with the carrier to make sure you buy the appropriate usage package and ISDN service (residence versus business, for now, although we believe that the business/residence classification will move to more flexible usage packages).

The basic design of ISDN is as a switched service for dialing different locations. It was not designed for connecting to one location and maintaining a full-time connection. There are many other network communication options that were designed specifically for permanent connections.

This leads to the ideal use of ISDN for RLA or LAN-to-LAN connectivity. It is best used for frequent, short connections to a LAN. It also is suitable for small office connectivity and leased-line backup connections, which are infrequent but cost-effective for these sites. When the backup ISDN line or the frequency increases to nearly full-time use, it is time to consider other network communication options.

TIP

Shiva Corporation (www.shiva.com) has developed an excellent "white paper" on bandwidth management, which describes some tariff issues and techinques for managing your ISDN equipment and network usage.

You may be able to reduce usage costs by using intelligent software that can automatically connect and disconnect with the LAN system as needed. Many equipment providers offer software that dynamically controls the line for the end users. We'll cover this topic in more detail in Chapter 11.

If you are planning to use ISDN connections for RLA applications on a large scale, you may want to pursue a contract rate with your carrier. This could increase the number of users eligible and lower your standard rates. Some carriers may offer special usage plans if you generate enough connections. If you cannot find a good flat-rate option, and the residence rate is higher than the business rate for your RLA application, consider a business usage contract for all of your users that can be discounted with your other business usage for each carrier.

PART III

How Do I Get
Connected?

CHAPTER

NINE

9

How Do I
Get Started?

- ■ How to use your ISDN worksheet

- ■ Criteria for deciding if you need ISDN

- ■ How to check ISDN availability in your location

- ■ ISDN hardware and software requirements

- ■ How to order ISDN from a carrier

The chapters in Parts 1 and 2 of this book have explained how technological change and the general use of technology have driven the need for larger bandwidth and faster network connections. ISDN service provides the high-speed connections for today's communications.

This chapter is designed to get you started with ISDN service and to furnish you with practical information that will lead you through the installation process successfully. As mentioned in earlier chapters, new standards are evolving as vendors make changes and improvements to their ISDN equipment. Unfortunately, until ISDN and ISDN equipment become more user-friendly or integrated, the user needs to understand the process of getting connected to make a successful transition from the analog world into the digital world of ISDN.

An Overview: Working through Your ISDN Worksheet

Figure 9.1 shows the summary version of our ISDN worksheet, which you will be completing as you work through Part 3 of this book. The ISDN worksheet is designed to help you ask better questions and guide you through the ISDN provisioning process, walking through the steps in the correct sequence.

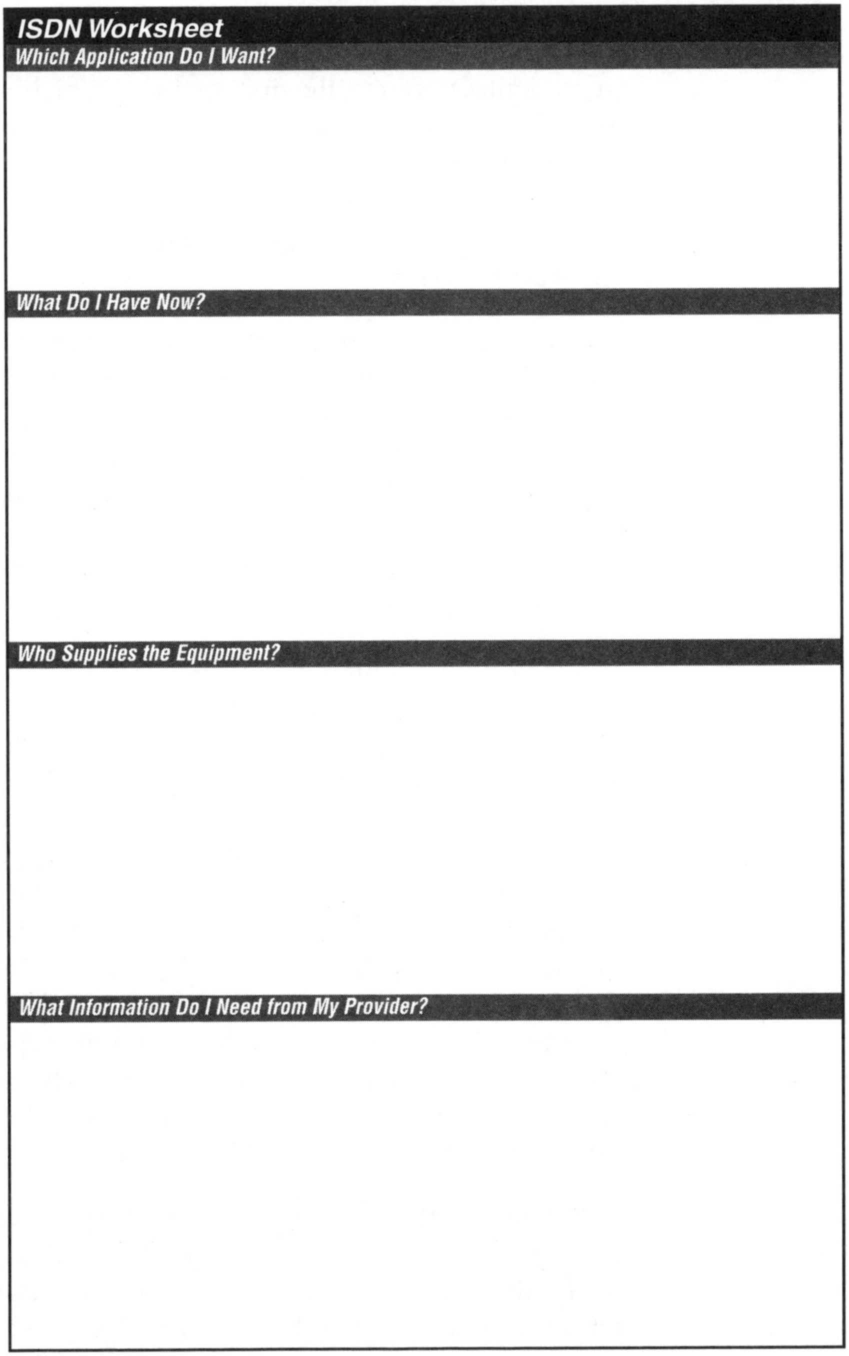

FIGURE 9.1:

ISDN worksheet,
summary version

ISDN Worksheet

Which Application Do I Want?

What Do I Have Now?

Who Supplies the Equipment?

What Information Do I Need from My Provider?

The worksheet shown in this chapter is a summary of the areas or questions you will need to answer to help make your ISDN setup successful. Of course, the first section, "Which Application Do I Want?" is the most important. We want to emphasize that the "secret" to implementing ISDN is to work backward, or to begin at the end. This is no different than basic goal setting. The goal is what you expect to be doing with ISDN service after the service and equipment are installed.

After you decide on an application, you may want to skip forward to the chapter covering your chosen application. Of course, you are welcome to read the others, but we have designed Part 3 to get you started as quickly and easily as possible. The chapters in this part are devoted to several specific uses of ISDN service, the related equipment, and support. Each chapter in this part includes a customized version of the ISDN worksheet to accommodate the application's special requirements.

After you pick your application and you have a direction or goal, you can begin shopping for a vendor and equipment. But wait, it's time for the "What Do I Have Now?" section of the worksheet. We recommend that you take an inventory of what you've got. As with any type of goal setting, you need to know where you are starting from before you can determine where you are going. You will need to know things like the type of your computer (for example, IBM PC compatible or Macintosh), the speed of the CPU (for example, Pentium or Cyrix, 586 or 486, 100 MHz or 166 MHz), the amount of RAM, and the amount of storage space (available or free hard disk storage). Also, investigate the communication port configuration on your PC and the hardware connected to each port. The ISDN worksheets in the chapters that follow should help you to keep all the information documented and handy for future reference.

With information about your current equipment in hand, you are ready to start looking for a vendor. This is where the "Who Supplies the Equipment?" section of your worksheet applies. As you call vendors, we bet you will find the information you gathered on your current equipment invaluable. Depending on your application, you will need to share certain details about your current equipment to help the vendor provide you with a product that best suits your needs. The information you've gathered will also be useful when you call your ISDN service provider or ask for help from one of the many support centers.

For example, if you are considering a videoconferencing application, you will need to know the speed and type of CPU and the amount of RAM. This information will help the vendor's representatives to determine if their "add-on" equipment will be compatible. Also, configuration information helps the vendor make specific recommendations or suggestions to help you upgrade your current equipment configuration, hardware, and/or software, if required. You may want to upgrade your current equipment or accept trade-offs between your computer's CPU performance or the amount of RAM inside your machine. In addition, you will need to consider choices to purchase external hardware devices or integrated (internal) hardware that must be installed inside your PC. All of these decisions will be covered in greater detail as they relate to your chosen application in the chapters that follow.

Next, you will return to the ISDN worksheet to record the new information and proceed to the next section, "What Information Do I Need From My Provider?". Like a locomotive gaining steam, the information you have gathered in the previous steps will be used to help you answer questions and request the proper provisioning from your local service provider. Depending on your individual situation, some portion of the information will be requested by your ISDN phone company or carrier. Ideally, the ISDN worksheet will guide you through the proper steps and help you keep track of details that are often overlooked or omitted in the ISDN implementation process.

But before you begin your journey, we suggest you take a minute to verify that ISDN is the "network solution" for you.

Are You Ready for ISDN?

Maybe you are reading this book because your friend has ISDN and shared his success with you. Now you are trying to determine if ISDN is the best choice for your situation. The following sections will help you make that decision.

Does the Application Need ISDN?

The main question to ask yourself is "Does the application really require ISDN?" This may sound simple, but as with any evolving technology, you need to separate fact from fiction by evaluating the following ISDN attributes in relation to your application needs:

- Dial capability

- Any-to-any characteristics of ISDN

- The 56/64 kbps capacity with one B channel

- The 112/128 kbps capacity with both B channels

You need to make sure that you are not trying to make ISDN perform unnatural acts for your application or environment. For example, do you need a dedicated line, or is a dial-up operation sufficient?

As an example, suppose you are considering ISDN for a network LAN link that needs to be up all the time. It may be an ideal private line (leased line) or frame relay circuit application, but it's a poor use of ISDN. It may also cost you many times more than a leased line. Both usage cost and the distance between network nodes must be determined to make a "smart" comparison.

Many people use ISDN for certain applications simply because it is inexpensive (or free) compared with other options such as frame relay. For example, Education First is a Pacific Bell initiative to connect schools and libraries using ISDN. ISDN is being implemented by schools because it is a free service and it doesn't need to be ordered. In many cases, they use the line for a permanent connection, and a permanent connection is not the best application of ISDN. Some customers have contracts that include flat-rated ISDN. Again, this may be cost-effective, but it still does not exemplify the ideal condition for ISDN applications.

Similarly, with ISDN Centrex, which can be a local-usage-capped environment, ISDN lines are used for Web servers with permanent connections because Centrex calls to the Internet service provider's can be free for local users. Again, this is not a good use for the ISDN service "dial" capabilities.

Are You Ready for New Technology?

Are you ready to take on new technology? Many people or companies believe they are ready for new technology, but underestimate the conversion time and/or the required resources. Unfortunately, new ISDN users believe that implementing ISDN is like turning on a switch. Fortunately, ISDN is getting easier to purchase and to implement. Still, while stepping through a full equipment and service installation cycle, someone will need to invest some time into the conversion process. The involvement will demand more than performing a PC configuration or making a single call to the local carrier. You will need to prepare and plan for ISDN. This may involve some or all of the following:

- Installing a new line and phone jack (possibly separate from your existing service or phone jack)

- Adding an NT1 device (which may be integrated in the terminal adapter)

- Purchasing new terminal adapter equipment for your PC or terminal (which may be external or internal to your PC)

- Wiring new cabling to accommodate your existing system

None of these are difficult, but you must be prepared to manage and accommodate the changes ISDN will require. Along with the adjustment in mindset, which applies to most changes in new technology, you will need to change certain activities to conform with the ISDN implementation for you or your organization.

Verifying Availability

If ISDN is the right choice for you, the next question you need to ask is "Is ISDN available where I need it?" Availability sounds obvious, but for ISDN service availability can be either highly probable or very improbable based on your location(s) and your local provider.

Imagine buying equipment, installing the equipment, calling your local access provider, and discovering that ISDN is not available in your area. Or you may have ISDN and desktop videoconferencing installed at your organization's location and want to connect with a client or branch office in another state. Unfortunately, your client or branch may not have access to ISDN service, or the cost may be prohibitive.

Although the overall availability of ISDN is high and increasing, especially in metropolitan areas, make sure that your users or community of interested users (friends, customers, or clients) are able to get ISDN cost-effectively and in a reasonable time frame. If most of your users do not have any access to the service or their access is limited, you won't see much improvement in communications for the whole body or community of users. You may invest a lot of your time and effort trying to convert a small number of users for very little benefit.

Table 9.1 shows some of the ISDN information hotline phone numbers for the RBOCs. You'll find more contact names and numbers for various carriers in Appendix A. In addition, the World Wide Web Home Page Reference list for various carriers contains tools that can help you find numbers and exchanges for service availability, determine current pricing, and access support staff.

TABLE 9.1:   ISDN Information Hotlines

RBOC	HOTLINES*
Ameritech	1-800-832-6328
Bell Atlantic	1-800-570-4736
Bell South	1-800-428-4736
NYNEX	1-800-438-4736
Pacific Bell	1-800-472-4736
Southwestern Bell	1-800-792-4736
US West	1-800-246-4736

* 4736 = ISDN

Availability by Region and by Company

Figure 9.2 is a visual index that you can use to determine nationwide availability of ISDN. Table 9.2 summarizes the availability by RBOC and region.

FIGURE 9.2:

ISDN worksheet, summary version

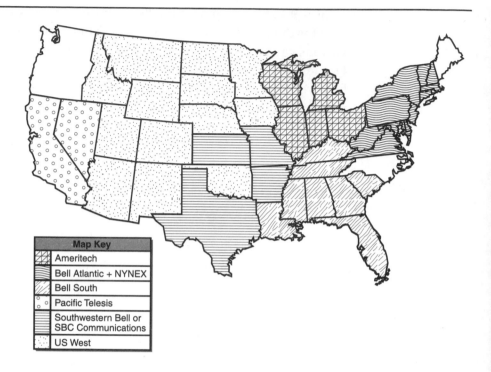

TABLE 9.2:   ISDN Availability by Company and by Region

ISDN Provider	Availability (by Region)
Ameritech	IL, IN, MI, OH, WI
Bell Atlantic	DE, MD, NJ, PA, VA, WV
Bell South	AL, FL, GA, KY, LA, MS, NC, SC, TN
GTE	CA, FL, HI, ID, IL, IN, KY, NC, OH, OR, PA, TX, VA, WA
NYNEX	MA, ME, NH, NY, RI, VT
Pacific Bell	CA, NV (Northern)
SNET	CT

TABLE 9.2:   ISDN Availability by Company and by Region (continued)

ISDN Provider	Availability (by Region)
Southwestern Bell (SBC Communications)	AK, KS, MO, TX
Sprint (Centel)	National (and local)
US West	AZ, CO, IA, ID, MN, MT, ND, NE, OK, OR, SD, UT, WA, WY

NOTE   Two mergers currently in progress may affect the information shown in Table 9.2. Pacific Bell's and Southwestern Bell's parent companies have announced (in April 1996) a merger, as have NYNEX and Bell Atlantic. It is anticipated that by the end of 1996, these mergers should have been completed. Look for updates on the respective World Wide Web home pages of these companies for related announcements.

ISDN Cost Variations

You also will want to consider the costs involved with implementing and using ISDN. Table 9.3 provides an overview of the variability of installation and costs in different regions or by different carriers.

TABLE 9.3:   ISDN Installation, Monthly, and Usage Costs

ISDN Provider	Service Name	Installation Charge	Monthly Fee	Usage (per min.)
Ameritech	ISDN Direct	$135	$34	3 to 9 cents
Bell Atlantic	ISDN Anywhere Program	$125	$28	1 to 2 cents
Bell South	Single-line ISDN	$200	$58 to $70, one or two B channels	None
GTE	Single-line ISDN	$80 to $150	$30 to $50	3 cents per B channel

TABLE 9.3:   ISDN Installation, Monthly, and Usage Costs (continued)

ISDN Provider	Service Name	Installation Charge	Monthly Fee	Usage (per min.)
NYNEX	Basic Rate ISDN	$230	$24	
Pacific Bell	Basic Rate ISDN	$125 (1 year) $ 34.75 (2 years)*	$24.50	1 to 5 cents
SNET	Digital Enhancer	$265	$50	15 to 35 cents
Southwestern Bell	DigiLine	$349.60 $224.60 (1 year)* $99.60 (2 years)*	KS & MO: $57.50 to $104.50 TX: $58 to $156	KS & MO: 0 to 4 cents TX: None
Sprint	ISDN Solution Center	$100 per B channel	Not available	4 cents per B channel
US West	Basic Rate ISDN	$110	$39 to $60	2 to 7 cents per B channel

*With this term of commitment to ISDN service.

The carriers are now developing a flat rate or tiered usage packages for ISDN service. Watch for these developments; you will see many new rates available by year-end.

Remember, ISDN tariff conditions and prices change frequently. The cost comparison that you perform today probably should be redone every six months to make sure that you know what the costs will be for your ISDN monthly fee and usage charges. Use the references and home page list provided in this book to keep up-to-date and learn about complete service conditions or limitations. Many companies have different tariffs (prices) for each state, and these tariffs will vary.

In some cases, you will have the option of ordering one or two B channels. Your decision will be included on your ISDN worksheet, based on your application requirements. Always verify the costs of ordering one or two B channels at the time of installation. Some carriers may charge for each B channel separately. In other cases, you pay nothing to add the second B channel at the time of installation. If your ISDN equipment supports a second B channel, go ahead and have it installed.

Line Extension Technology

As we've explained in earlier chapters, your phone company must have the necessary equipment in the central office in order to provide ISDN service. Some phone companies offer what is called ISDN Anywhere service, which means if you order ISDN, they will find a way to get it to you.

In cases where the phone company does not have the right equipment in the local central office that serves you, the company can use line extensions or extender technology to serve you from another adjacent central office. Unfortunately, the use of line extension technology may significantly increase the cost of your ISDN service. You may also be assigned different phone number prefixes (the first three numbers that follow the area code) than you had before ISDN.

As a rule, ISDN is more available in metropolitan, urban, and suburban areas and more difficult to get in rural or remote areas. In some locations, the extender technology will not work, or the existing copper wires are not "clean" enough to provide ISDN service. Keep in mind that even analog phone service does not enjoy 100 percent availability in the United States.

RBOC Deployment

Installation of ISDN equipment in central offices has been slow, but a significant number are now capable of providing ISDN service. Figure 9.3 shows a graph that summarizes ISDN lines in the United States.

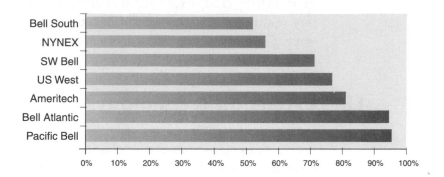

FIGURE 9.3:

ISDN-capable RBOC lines nationwide

Pacific Bell is the leader among RBOCs, with about 95 percent deployment. According to Pacific Bell, in the San Francisco Bay Area, nearly everyone can get ISDN, and the few pockets remaining without service will have it within a few months. Bell South is behind, with roughly 50 percent deployment. Most major metropolitan centers are well covered, but rural sites are not. GTE has taken a less aggressive ISDN position, but the company is installing an ISDN overlay network and will provide ISDN-to-analog wire centers in the interim.

Other ISDN Providers

You can get ISDN service from companies other than the RBOCs mentioned earlier. Table 9.4 lists some additional providers in the United States and other countries.

TABLE 9.4:   Other ISDN Service Providers

Location	Provider
United States	AT&T Digital Long Distance Service Northern Arkansas Telephone Co. Nevada Bell Sprint Digital Long Distance Service WilTel
Canada	Stentor: The Alliance of Canada's Telephone Companies CanISDN: The Canadian ISDN Resource Centre

TABLE 9.4:   Other ISDN Service Providers (continued)

Location	Provider
Europe	Deutsche Telekom British Telecom France Telecom Helsinki Telephone Company Swiss Telecom Tele Danmark
Israel	BEZEQ, The Israel Telecommunication Corp. Ltd.
Australia	Telstra
Japan	Nippon Telegraph and Telephone Corp.
Africa	Telkom S.A.

How the Phone Company Checks Availability

In some cases, your local phone number and address are all the phone company needs to determine the availability for your area. However, an ISDN service provider may need to check two items.

First, the phone company may need to see if there are a sufficient number of twisted pairs, or copper wires, to provide ISDN in your area. In some older areas, your local phone company may not have the two "working" pairs needed to hook up ISDN. Why two? Standard Basic Rate ISDN requires two pairs: one for each B channel. Also, as discussed in Chapter 2, ISDN equipment relies on an external power source—your home or office's electrical outlet. If you want a phone line during a power outage, you'll need to keep your analog service. Keep twisted pair for your analog service. In some cases, the phone company will need to dispatch a crew to your area to repair lines so at least two additional twisted pairs can be installed at your location.

Second, depending on your distance from the phone company's central switching office, a crew may need to perform a "loop qualification" test to see if the distance between your location and the ISDN switch is within this limit. Loop qualification is a test to check the physical distance and the quality of the line between your connection and the central switch or central office. If the distance exceeds the 18,000 foot limit, the phone company may need to send out a crew to go under the street in your neighborhood and install a signal amplifier or repeater to boost or strengthen the signal. This limitation is based on a 26-gauge twisted pair. Your location may have a larger gauge, making it possible to exceed the 18,000 foot limit.

If the phone company needs to check either of these items, you may need to wait from three weeks to three months (or more) before it can establish ISDN service to your location. At some phone companies, the representative may tell you that it will take three weeks or more just to schedule the loop qualification test. If the test shows special equipment is needed, your local phone company may need additional time to schedule and dispatch a work crew to install the equipment in your neighborhood.

On the other hand, you may not need to have any more wires running into your house; ISDN service won't require digging up your front yard. In other cases, you will be able to add a third line directly to your den or home office, allowing you to use ISDN as the third and fourth line for your business phone, fax, and online service access, without disturbing the other two home lines.

Know Your Needs

The business requirements that lead to ISDN are fairly explicit. The most common one is a need for high-speed, low-cost, switched circuits for branch offices within a relatively small geographic area. ISDN is also often considered a *dial-backup* or *dial-on-demand* technology, especially when redundancy or responsiveness for mission-critical applications is a concern.

Long-distance ISDN is often used for videoconferencing. *Telecommuting* is an additional incentive for ISDN adoption because of the technology's ability to provide home-based users with high-speed (64 kbps or greater) connections and POTS services.

Setting up your ISDN service requires that you take responsibility to translate communication requirements into their ISDN equivalents. This translation process requires a thorough understanding of the related features, options, and nomenclature.

ISDN's rich feature set and the residual proprietary nature of vendors' switches require that you take a different approach than you do when dealing with conventional leased-line-based networks. The first step is to gain a thorough understanding of the technical communications requirements involved, including those related to call features, capacities, and physical interfaces for voice, data, and video.

It is imperative to have a clear sense of the communications tasks that need to be accomplished. Remember, what is the end state of your ISDN activities or application? Remote LAN access is not a communications task but rather a business solution.

ISDN can provide a myriad of connection options, and chances are you'll receive every available feature unless you're very clear about what you want to achieve with ISDN. You may not need or want to pay for these additional features.

A Few More Considerations

The benefits you can enjoy by switching to ISDN service depend on why you are considering the service and how you would like to use it. Many residential customers are attracted to the fact that they can replace two regular analog lines with one ISDN line. Businesses appreciate the high-speed, reliable connections provided by ISDN service. However, there are several questions you need to ask yourself before proceeding further.

Do You Control Your Community of Interest?

If you will use ISDN to connect your home PC to your office computer, you have control over the type of equipment that you use at both ends of the communications path. In this case, you can use equipment that transmits digitally at 64 or 128 kbps. On the other hand, if your goal is to access a public service such as the Internet, you'll need to ensure that your equipment is compatible with the equipment of your Internet service provider (ISP). The main categories of ISDN rate adapted access methods are asynchronous V.120, synchronous and asynchronous V.110, PPP, and Multilink PPP.

You may also need to continue to use your existing modem and fax equipment to communicate with people in other companies where you can't control the selection of equipment. In this case, you'll need an ISDN terminal adapter with analog conversion ports where you can connect the analog phone, modem, and fax machine that you already own. This should always be a consideration—few of us want to abandon or throw away existing home office equipment.

How Big Are the Files You Intend to Transfer?

Macintosh and Microsoft Windows users have access to many files and applications other than text. Voice, music, still images and video clips may require faster transfer speeds than are possible with even the fastest modem.

With ISDN, the starting point is a single 64-kbps B channel. However, equipment is available that can combine any number of B channels to create even higher transmission rates.

Do You Want Advanced Voice Features?

Some ISDN phone equipment provides advanced voice features that you would need to pay extra for if you had a regular analog line. With ISDN, depending on the equipment, you can have the following voice features:

- Put multiple calls on hold.

- Transfer calls to other numbers.

- Create conference calls.

- Forward all calls to a number you specify.

- Display the number of the person calling (even if you're on the phone).

- Block calls from a specified number, or automatically call back the last person who called.

Currently, most companies offer the newest caller-identification features, which will allow you to create your own personal call center in your home office. However, you need to have the software or equipment to manage these features and make them productivity tools for your work at home. With ISDN service comes the ability to program multiple phone numbers on B channels. This offers the potential of creating a very sophisticated call center in your home. You will see many new equipment packages utilizing these features soon.

ISDN Equipment Requirements

ISDN is the most versatile, feature-rich service ever offered by your phone company. However, the fact that it is so versatile and flexible means that there are many, many options that need to be specified when you first order your service. These options will vary depending on the equipment you plan to use on your line.

TIP Even if you plan to use your line only for data, order both B channels to support voice and data. If the extra B channel doesn't cost any more, and if you add an ISDN phone later, you won't need to pay another order-processing fee to the phone company.

What Hardware Do You Need?

The hardware you may need includes the network terminating device (NT1), a terminal adapter (TA), ISDN telephones, and an Ethernet ISDN bridge/router.

The NT1

The NT1 device provides the physical and electrical termination of the twisted-pair wiring coming from the phone company's central office. The NT1 also converts the two-wire twisted-pair connection into an eight-wire distribution system within your premises. This provides a standard interface for ISDN terminal equipment. Any terminal equipment designed to meet the ISDN standard can plug into this interface by using a standard 8-pin RJ-45 connector. The NT1 also provides remote diagnostic capabilities to allow central office personnel to perform centralized remote fault isolation.

The TA

If you wish to use your existing analog modems and fax machines on an ISDN line, you'll need some sort of TA that can convert from the analog non-ISDN protocol, or signaling method, into the ISDN interface. The TA performs the function of converting the existing analog interfaces into digital ISDN interfaces.

If you're planning to use ISDN for data, some ISDN TAs have the NT1 built in. This means a simpler and more reliable installation. In addition, some TAs have an additional S/T port into which you can plug more ISDN voice or data terminals. In the case of voice sets, the built-in NT1 means that you can only connect one set to the ISDN line. We don't recommend the latter unless you feel that the set has all the features you need for the foreseeable future. You never know when you might want to add another piece of equipment to your line.

The ISDN Phone

The major advantage of using ISDN phones is the ability to control multiple calls using a separate button for each call. The advanced features that are available on an expensive business phone system can now be enjoyed by the small business or residential subscriber. For example, using an AT&T 5ESS switch with two B channels, you can have up to 64 different phone numbers appear on a single ISDN telephone set.

Another option is to have several buttons assigned to the same number. In this case, if you're talking to someone on one button and another person calls you, the new caller won't get a busy signal. Instead, another button will flash, your phone will ring, and the number of the caller will be displayed on your set. You can then decide how to handle the call:

- Put the first call on hold to answer the new call.

- Send the new call to voice mail or a fax machine.

- Ignore it, store the number, and call the person back later.

- Forward the call to your office or another person's phone number.

New ISDN equipment will allow you to program many ways to manage your calls.

The Ethernet ISDN Bridge/Router

If you need to connect to an office computer on an Ethernet LAN, you may want to get an ISDN/Ethernet bridge/router. These provide remote workgroups and individuals with high-speed, digital, transparent access to distant enterprise networks and other resources, with all the services of their office when and where they need it using ISDN.

ISDN/Ethernet bridge/routers can combine both ISDN B channels and apply compression to provide data-transfer rates of up to 500 kbps over a single ISDN line. Most products route IP and IPX and bridge other protocols. In some cases, bridges are a better solution than routers. Since bridges perform MAC (media-access control) layer bridging, they will operate with any network operating system.

NOTE The ISDN calls are originated automatically by the bridge without complicated user procedures. The bridge sets up the first B channel when needed and drops it after a predetermined period of no activity. The second B-channel call is originated when the traffic reaches a predetermined threshold and drops when traffic falls below a set threshold for a specified period of time. Be sure to check how your ISDN provider charges for usage. In cases where the initial minute is higher in cost, this type of configuration can be expensive. See Chapter 11, which covers remote LAN access and telecommuting, for more information about LAN equipment for ISDN.

What Software Do You Need?

The software you may need for your ISDN setup includes COM port emulation, network connections, data compression, and BONDING.

COM Port Emulation

Terminal adapters often come with software that emulates a COM port to allow existing communications/modem applications to continue to work without additional, specialized software. To date, COM port emulation has been popular because it uses existing communications software, but the use of remote node technology and protocols is gaining industry momentum.

NOTE Some applications may require additional communication port hardware, which may not come with the ISDN hardware or software within an ISDN package or "bundle."

Network Connections

Other systems use popular protocols, such as SLIP or PPP, to connect to the network. These protocols allow the remote user to become a network node, with the computer thinking that it is directly connected to the LAN. Remote node connections remove the limitations of transferring files using a modem and terminal software.

Data Compression

Driven by the bridge/router and remote node vendors, a de facto compression standard has emerged. A derivative of Lempel Ziv (LZ) is the compression protocol of choice for ISDN. Effective for compression at high data rates, LZ is considered more effective than V.42bis (commonly found on most analog modems). With LZ, 2.5:1 compression rates are common, with rates as high as 4:1 possible.

NOTE

In applications that access the Internet, compression may not gain much of an advantage. You will find that most image files, program files, and movie or video files are already stored as compressed files. Additional compression using your ISDN equipment will have little to no effect.

BONDING

When it comes to ISDN services, most of today's Information Services groups will be concerned with BRI circuits, due in part to the flexibility offered by their channels.

The two B channels in a BRI circuit can be bonded at both ends of the circuit, providing a switched-circuit data path for data or video-conferencing applications. If both a telephone circuit and a data circuit are needed, one B channel could be configured for voice and the other for data.

Using such options, the Internet can be accessed at 64 kbps, or circuit BONDING can be used to provide a 128-kbps circuit for medium-quality videoconferencing. Credit-card verification on the D channel could occur at any point during data and voice transactions.

NOTE

BONDING refers to inverse multiplexing, as defined by the Bandwidth-on-Demand Interoperability Group. See Chapter 3 for details.

The flexibility doesn't stop there. A remote site can be configured with a router that has a 128-kbps switched-circuit interface to the home office. This interface connects on demand and disconnects during inactive periods. The remote site gets a high-speed link at a fraction of the cost of a fractional T1 circuit.

In another scenario, a telecommuter could have an analog phone and fax machine connected through one B channel, and an ISDN NIC (network interface card) and NDIS (Network Driver Interface Specification) and PPP (Point-to-Point Protocol) software connected via the other B channel to the home office.

Other options include a router connected to the Internet via one B channel. This site might use the second B channel to emulate a direct inward dialing (DID) circuit for a lab or to connect to a remote access server dial-in/dial-out analog circuit.

Crunching the Numbers

The next task is to determine the total number of each of the following elements that will be required:

- Non-ISDN equipment (TE2) and the corresponding interfaces (RJ-11, RS-232, and V.35)

- ISDN equipment (TE1) that connects to the S/T interface circuit-switched voice and data channels

Your organization must also determine the function of the D channel: signaling only or signaling and packet-switched data. This helps determine what equipment will be needed and the desired method by which the ISDN service provider provisions the circuit.

A Pop Quiz for Your ISDN Equipment Vendor

Vendors have found significant marketing opportunities in the ISDN interface arena. The quantity and variety of devices have burgeoned, creating much confusion over the potential combinations. Ask your equipment provider the following questions when categorizing the technical requirements of ISDN products:

- Does the product support more than one B channel? Some devices support only one B channel or have a second channel that can only be configured for voice.

- Does the two-data-mode, B-channel product support the BOND-ING standard? Does it support compression? Can you receive a call while busy on the line?

- Does the product provide analog phone support (a POTS interface)? How many lines?

- Is battery backup available? How long does it operate? (Unlike analog phone service, ISDN phone service doesn't provide power for your equipment.)

- What type of ISDN connection is provided? If you purchase equipment with a U interface but don't have an S/T or a TA interface, you're stuck with a single-purpose device. Try to find equipment that provides you with an external S/T interface so you can expand in the future.

- Does the product support IP and PPP for Internet connections?

- Does it provide V.120 capability (the prevalent data rate adaptation protocol in North America)? Do the V.120 or V.110 protocols apply to both synchronous and asynchronous services?

- Does the product have an SNMP (Simple Network Management Protocol) agent and MIB (Management Information Base) for SNMP management?

- Does the product support your API (Application Program Interface) of choice, such as Telephony API (TAPI) or WinISDN?

- Does it provide enough flexibility for your TA connections, such as RS-232, RJ-11/RJ-45, RS-422, or V.35?

NOTE Most RS standards are now EIA (Electronic Industries Association), such as EIA-232-D. EIA is a trade organization of manufacturers that sets standards for use by its members and lobbies Congress on their behalf. The group is based in Washington, D.C.

- If your network configuration is many remote sites to one regional or central site, what is required at the host or hub site?

Don't forget to ask any additional questions that may be appropriate to your specific application. It's also a good idea to read all available product literature on candidate equipment and, if possible, obtain an installation manual before you buy the product.

> **NOTE**
> Interoperability is a concern with ISDN service. The lack of national standards has been an obstacle to widespread ISDN deployment. The key question is whether or not equipment that works on one ISDN system can function properly on an ISDN system from another source. Fortunately, much progress has been made in the area of standards development and compliance. Also, ISDN service providers are now striving to help customers identify their application or use of ISDN and equipment, so that they may provide better service and support.

Configuration for ISDN Service

In addition to the configuration the phone companies must do at their end of your ISDN line, there is also some configuration you must do at your end. Your phone company will supply four pieces of information that you need to know to make your ISDN service work with your desktop or Windows-based PC unit:

- Phone numbers (sometimes called DNs for directory numbers)
- Switch type (usually 5ESS or DMS)
- SPIDs (Service Profile Identifiers)
- ISDN version (National ISDN or Custom ISDN)

What's Your Phone Number?

The first type of information you need is your phone number or numbers. In some cases, each B channel on an ISDN line has its own number; in other cases, both B channels share a single phone number. In part, this depends on the switch type in your central office.

Your phone company will tell you how many numbers your ISDN line will have. Separate numbers may be useful if you plan to take incoming calls on your ISDN line.

What's Your Switch Type?

Another piece of information you'll need to get from your phone company or ISDN provider is the switch type. The configuration software for most ISDN terminal adapters needs the connection's switch type.

The switch type simply refers to the brand of equipment and software revision level that the phone company uses to provide you with ISDN service from the central office that supplies your area. Worldwide, only a few switch types are used by carriers to provide ISDN. Some countries use only a single switch type, but the United States and its carriers use several. AT&T 5ESS, the Northern Telecom DMS-100, and the Seimens EWSD are the three most popular switch types in North America.

During the last ten years, most carriers have been deploying or installing the 5ESS or the DMS100 switch. All RBOCs use those two and most deploy them somewhat evenly. Only three have deployed the Siemens switches in any scale or volume.

AT&T 5ESS Switch

Of the new switches being installed over the past two years, the 5ESS represents about 60 to 66 percent. All the new competitive access providers (CAPs), who are installing new switches in major cities across the country, are deploying 5ESS switches. What does this means to you? In the near future, ISDN service off the 5ESS switch will be highly probable.

NOTE The AT&T 5ESS is now called the Lucent 5ESS as a result of the recent divestiture of AT&T business and research companies.

The 5ESS switch has the ability to support eight terminals on one ISDN line. This is uniquely different than the European implementation, which assumes only one terminal per ISDN line (and therefore does not need the SPID convention, described shortly).

DMS-100 Switch

The "other type" of phone company switch, the DMS-100, was the first switch to implement ISDN. It was designed during the early days of ISDN and did not comply with the standards relating to SPIDs. (In fact, the standards in use today were not even started when the DMS-100 was designed.) The early implementation did not have the robustness for high-volume ISDN lines or traffic. Although it has been redesigned since the early days, the switch still needs further improvements.

The DMS-100 switch, based on early standards, assigns one SPID to each B channel, rather than one to each device. Therefore, if your nearest central office's switch is a DMS-100, you will be limited to two ISDN devices connected to your customer premises equipment (CPE), rather than eight.

What's Your SPID?

The SPID was developed by Bellcore to help identify and create a service definition for individual terminals on individual ISDN lines. Once multiple terminals were allowed on one ISDN line, there had to be a mechanism for identifying the features (service profile) for each terminal on the same line.

As explained in Chapter 3, SPIDs, short for Service Profile Identifiers, are used in only the United States and Canada. The SPID usually consists of prefix digits (almost always two, such as 01) and the phone number with some additional suffix digits (usually three or four, such as 0000) added to the beginning and/or the end. SPIDs are sent by your ISDN equipment to the phone company's central office or switch. The process informs the switch which service profile each terminal will have and what (outgoing call, features for each phone number, data or voice equipment, and so on) to expect. The phone company must tell you the SPID that is associated with your ISDN numbers and lines.

This identifier helps the switch understand what kind of terminal or equipment is attached to the line and the related features for each terminal on the ISDN line. If there are multiple devices attached, the SPID helps route calls to the appropriate terminal or device on the same ISDN line.

NOTE

Another plan for the SPID was to use it to allow users to move terminals around a Centrex ISDN system. The idea was that the terminal could signal by identifying itself by terminal type and associated SPID number. In this way, the switch could simply configure the line for that terminal by using the SPID configuration of features. However, in some preliminary tests, there were major conflicts when the terminal moved from one location to another and an error occurred, so that neither transaction was completed. There was no way to easily identify what was the original location and what was the destination of the move when this error happened during processing. If the automatic reconfiguration failed, the ISDN service would not work while the problem was corrected. Within large organizations where this feature might be used, many telecommunication department managers had enough inventory problems without adding this to their workload. These problems explain why automatic configuration was never really adopted by large companies, who were the intended customers.

If you are hooking up only a single device to your ISDN service (for example, you're setting it up in a point-to-point configuration), you might not need a SPID at all, because the phone company can identify your ISDN line as one particular type, full time. This depends on what equipment the phone company has in the central office; the DMS-100 switch still requires that you have a SPID for each B channel.

Currently, there is a major movement in the telephone and terminal adapter industries to simplify and standardize the SPIDs for all the phone companies providing ISDN and vendors supplying customer terminal equipment. In fact, Adtran announced an ISDN equipment feature that is one of the first attempts to configure ISDN equipment with the SPID numbers, without involving the user. The software cycles through the most common SPID number formats until one works. Although this method will work in the majority of cases, it is not "bulletproof" and does not succeed in all cases.

The NIUF (National ISDN Users Forum) and an equipment vendor group (including Ascend, IBM, and others) have both been working on a resolution that will lead to one standardized format that will allow the majority of all applications on equipment from different vendors.

NOTE National ISDN Users Forum (NIUF) is a name you will eventually run into while exploring ISDN. The NIUF organization is a hero for ISDN users. Sponsored by the United States Department of Commerce under the National Institute of Standards and Technology (NIST), the NIUF assists in the development of ISDN applications and educates users on ISDN services.

Typically, your terminal adapter equipment has a setting for only one of these types of line configurations. Some equipment has some flexibility to be modified (or configured) for different ISDN versions and switches. There are actually a standard set of combinations (configurations) defined for setting up BRI lines. These are called National

ISDN Interface Groups (NIIGs), so there will be a limited menu of offerings available.

What's Your ISDN Version?

The configuration software for your ISDN terminal adapter will also require you to specify the ISDN version supported (National ISDN versus Custom ISDN) by your local carrier.

As explained in Chapter 3, National ISDN is the standard developed by Bellcore to standardize ISDN services. Some carriers do not adhere to National ISDN exclusively. They offer a separate version, called Custom ISDN. Your carrier will indicate which type of ISDN service it supports.

The Formation of the Vendors ISDN Association

Twelve networking industry leaders—Adtran, Ascend Communications, Bay Networks, Cisco Systems, Digi International, Eicon Technology, Intel, Microsoft, Motorola, Shiva, 3Com, and US Robotics—have together formed the Vendors ISDN Association (VIA), a nonprofit, California-based corporation dedicated to making ISDN more accessible to businesses and individual users. The VIA expands the group initially known as the ISDN Forum, announced January 23, 1996. The formation of the Association includes formal affiliation with the National ISDN Users Forum (NIUF) and the National ISDN Council (NIC).

VIA's initial board of directors and officers, already in place, are guiding current activity, including drafting a VIA technical position on ISDN implementation. In addition, VIA operates committees to provide the appropriate forum for discussion and resolution of technical and marketing issues.

VIA participants will exchange technical information, address ISDN user needs, and promote interoperability for faster and easier digital transmission of voice, images, and data across copper telephone lines.

Membership in VIA is open to any interested organization, and members are actively encouraged to participate in committees and meetings. Currently, more than 90 companies worldwide have expressed interest in membership. VIA is already responding to these inquiries and adding member companies including CPE vendors, network infrastructure vendors, network service providers, information service providers, and application developers. The procedures adopted by VIA for information distribution and comment, and the actual conduct of meetings, ensure members are heard and issues discussed prior to adoption of a formal position. As a result, VIA members have a unique opportunity to participate in the development of evolving ISDN product and services.

The initial focus of VIA's work will concentrate on automated ISDN configuration capabilities. ISDN users currently need to configure their systems for ISDN by entering information about their telephone switch and SPID. VIA participants will recommend ISDN configuration capabilities to eliminate manual configuration, making end-user setup transparent by the end of 1996.

"Businesses and consumers are eager to deploy ISDN as a quick, cost-effective way to gain faster access to the Internet, telecommuting, and videoconferencing," said Deepak Kamlani, corporate secretary for VIA. "Improved use and deployment of ISDN facilitates more widespread use of the technology and we see the VIA playing an important role in that development and promotion. Our goal is simple: for users to unpack an ISDN terminal device, plug it in, and be able to use it."

VIA's charter includes the following key objectives:

- Simplify and accelerate the availability of interoperable ISDN customer equipment solutions based on the needs of end users worldwide.

continued on next page

- Drive the development of application market requirements, based on open standards and their consistent implementation in end-user products.

- Simplify and/or automate ISDN equipment configuration, operation, and management.

- Promote open network interoperability standards, testing, and uniform processes for ordering and implementing ISDN by end users.

Information requests regarding membership in VIA should be directed to Deepak Kamlani of Interprise Ventures at (510) 277-8110 or dkamlani@inventures.com.

Ordering ISDN Service

The last, and probably most difficult, step in the process is to order the ISDN circuit. While the industry has made great strides in simplifying the task of ordering a properly provisioned ISDN circuit, the process still isn't foolproof.

Knowing the Service Types

The key is to remember that there are three basic types of services:

- Bearer services
- Supplementary services
- Teleservices

It helps to have the three defined according to your desired network configuration and capabilities.

Bearer Services

Bearer services are the entities in the ISDN architecture that actually carry information. These services don't alter any of the information as it's transmitted over the ISDN circuit.

Bearer services are available in two modes:

- *Circuit mode* is similar to a telephone circuit; a connection is made, perhaps through many interconnecting points, and dedicated during the entire connection period.

- *Packet mode* consists of many packets of digital information, all of which could be traversing separate network paths to the other side of the connection. Packet mode is typically associated with X.25 or frame relay services.

The current ITU-T definition of ISDN specifies ten circuit-mode and three packet-mode bearer services.

There are three subcategories of bearer services: line sets, mode, and voice and data settings. The bearer service modes most likely to be used in your environment will vary depending upon the application. The following are the most common:

- Circuit-mode 64 kbps, unrestricted, 8 KHz, structured bearer service. This service is basically a clear channel or transparent service that can transmit any kind of data. The 8 KHz designation refers to a timing function in which the circuit provides synchronization for services, sparing the user's application the task of providing timing.

- Circuit-mode 2 x 64 kbps, unrestricted, 8 KHz, structured multi-use bearer service. This service is suitable for data and allows the two B channels to be combined in a way that solves problems associated with the timing of data that is sent over independent channels.

- ISDN virtual-circuit service, case B. This service is a packet-mode bearer service. (Case A is designed for X.25 service and isn't widely used.) With ISDN virtual-circuit service, the D channel can be used for packet data transfers. The service is slow (only 9,600 bps), but it's adequate for tasks such as credit-card verification.

Voice and data settings are specified by the user during the configuration and setup of the ISDN equipment. If the RBOC provisioning does not agree with your configuration settings, the ISDN terminal adapter will not synchronize with the switch (or turn on the little green lights, usually on the front of your ISDN equipment). Configuring the ISDN equipment to match the RBOC's switch provisioning is necessary to make a good connection.

Supplementary Services

Supplementary services compose the voice and telephone side of ISDN. These services provide functions such as call waiting, call forwarding, and caller ID. You may find that the wide variety of configuration parameters and service combinations offered makes supplementary services difficult to comprehend and select.

Teleservices

Teleservices comprise ISDN's gray area. The ITU-T defined most of these services with the European market in mind. Teleservices are enhanced telephone services such as telephony, videotext, telex, teletex, telefax, teleconferencing, video telephony (television over ISDN), 7 KHz audio (very high-quality audio), and message handling. Most of these services won't be a large concern in the ISDN arena for at least the next few years.

Contacting Your Carrier

The next step is to order the desired services from your local exchange carrier, which entails some knowledge of local tariffs, Bellcore standards, and nuances of the NIUF ordering standards. Basically, your task at this stage is to select the bearer, supplementary, and teleservices characteristics that you would like the provider to use to provision your ISDN circuit.

The Channels

The first step in the ordering process is to determine the number and type of channels or line sets required. This could mean, for example, a network with no B channels and one D channel, one B channel and one D channel, or two B channels and one D channel. The only restriction is that one D channel must be included to serve as the signaling and control channel.

The Operational Modes

The next step is to select the operational modes required for the B and D channels. For example, you might order BRI service with two B channels and one D channel, with circuit mode on one B channel, packet mode on the other B channel, and signaling and packet mode on the D channel. Add more configuration parameters for voice and/or data on the B channel, ISDN subscription parameters, and supplementary services (feature sets), plus any desired teleservices, and you've got a potential disaster from an ordering standpoint. Unfortunately, the more complex your order, the higher the potential for the RBOC's service technicians to make an error.

Tariffs

RBOCs are regulated, and their ISDN services are tariffed. The utilities commissions and the RBOCs negotiate features and their costs for each RBOC product offering.

Examine tariffs carefully to see which offering is most cost-effective for your environment. Do you really need voice and data on both B channels? One region's ISDN line set 17 (voice and data) might have a higher tariff than another region's line set 14 (data only), so it's crucial to verify these tariffs before ordering.

Simplifying with Ordering Codes

The NIUF and the ISDN Solutions Group have been working to simplify the ordering process. As mentioned in Chapter 3, these efforts have led to ISDN Ordering Codes (IOCs).

IOCs take the guesswork out of ordering ISDN service. An IOC is a product-specific ISDN ordering code assigned to the product by Bellcore and accepted by most phone companies. For example, *3ComA* is the IOC for the 3Com Impact digital modem. *IBM Blue* is the IOC for the Intel Proshare desktop videoconferencing product.

These IOCs are a combination of line sets, modes, feature sets, and supplementary services. A table of IOC nomenclature can be obtained from the NIUF on the World Wide Web at http://www.ocn.com/ocn/niuf.

Carrier-Supplied Information

After you've given the service provider the necessary information, request that the provider supply you with the information you'll need to configure your equipment.

Don't be alarmed, however, if the provider has never heard of ISDN IOCs or capability code. This just means you'll specify the line set, modes, and feature sets you need on your own. Most providers have standardized packages of line and feature sets, and you'll need to select the one that most closely matches your communications needs.

There are some designations you should be familiar with to simplify the ordering process. We have described the SPID earlier in this chapter and in Chapter 3. To summarize, the SPID is a unique number used by a particular piece of terminal equipment to define itself to a switch. A DMS-100 (a Northern Telecom telephone switch), for example, uses a SPID that's typically a 12-digit numerical code consisting of the user's area code, ISDN telephone number, and a two-digit identifier. See Chapter 14 for more information about SPID formats.

Your directory number is your telephone number. If you have a terminal adapter for an analog phone or ISDN phone, outside callers can contact you using your directory number.

An ISDN line can have multiple directory numbers. The terminal endpoint identifier (TEI) is used to uniquely identify ISDN devices, since up to eight such devices can be daisy-chained to the S/T reference point. The most common setting is automatic; the device and the switch negotiate which TEI is assigned during the session.

CHAPTER

TEN

10

A Guide to ISDN Setup for Online Information Access

- Selected ISDN terminal adapters

- What your ISP provides for ISDN service

- What you provide for ISDN service

- Information from your local exchange carrier

- Information about your ISP configuration

The World Wide Web and the explosion in Web applications, such as Java, VRML, and streaming video and audio, are putting a question in every Web surfer's mind: "How can I go faster?" Combine the need for speed with the increasing availability of ISDN service and you have a powerful combination: the World Wide Web, with an increasing variety of bandwidth-hungry applications, and ISDN, delivering the "right here, right now" bandwidth.

The ISDN worksheet shown in Figure 10.1 is provided to help you bring all the pieces together and avoid wrong turns and potholes as you merge onto the Information Highway's version of the express lane, ISDN. Ordering ISDN service can sometimes be confusing and frustrating. We've prepared this step-by-step guide to get you up and running with ISDN service.

Which Application Do I Want?

This chapter is intended for those individuals who want a high-speed connection to the Internet and want a dial-up connection rather than a dedicated connection. The worksheet and information here will help you make the transition from an analog to a digital connection using ISDN to access the Internet four times faster than the commonly used 28.8-kbps modem.

If you are considering ISDN for connecting yourself or your employees to your office network, as well as to the Internet, see Chapter 11. That chapter is devoted to using ISDN for telecommuting and remote LAN access applications.

FIGURE 10.1: ISDN online information access worksheet

ISDN WORKSHEET	Page 1 of 3

Which Application Do I Want ?

Online Services

What Do I Have Now?

Local ISDN Availability

☐ Is ISDN available from local phone company?

☐ Is a loop qualification needed?
Completion date ___/___/___

☐ Confirm installation charge $_____

ISDN Internet Access Availability

☐ Is there a local provider? (Yes > Skip to COM Port Availability)

☐☐ Is there a national provider?
If yes, is there a local point of presence?

COM Port Availability

☐ Is a serial port available?   COM 1 2 3 4

☐ Is the port a high-speed 16550 UART?
-Not sure?- See "Is It a High-Speed COM Port?" in text.

Who Supplies the Equipment?

What the ISP Provides

☐ ISP contact information
Contact name: _____
Company name: _____
Phone: _____

☐ Equipment recommendations?
A._____
B._____
C._____

ISP Shopping Parameters

☐ Basic monthly charge $_____
☐ Usage included in monthly charge
☐ B-channel minutes
☐ Per megabyte
-OR-
☐ Flat or fixed monthly rate
☐ Additional usage $_____
☐ Bundled services?
☐ Web page storage
☐ FTP Drop Box
☐ Other_____

What You Provide

☐ Need compression?

☐ Want MLPPP+?

ISDN WORKSHEET *continued*	Page 2 of 3

What Information Do I Need From My Provider?

Local Exchange Carrier

☐ Contact information
Phone company rep: _____
Company name: _____
Phone: _____

☐
☐ Installation date: _____
Service order number: _____

ISDN service charges

☐ Monthly charge $_____
☐ Per minute usage
    8 a.m. to 5 p.m. $_____
    from ____ to ____ $_____
☐ Additional charges $_____

☐ Installation information
Phone company rep: _____
Company name: _____
Phone: _____

Provisioning information

☐ Type of ISDN
  ☐ National
  ☐ Custom

☐ Digital switch type
  ☐ 5ESS
  ☐ DMS-100
  ☐ Siemens EWSD

☐ Local directory numbers
  ☐ B channel one   _____
  ☐ B channel two   _____

☐ Service profile identifiers (SPIDs)
  ☐ B channel one   _____
  ☐ B channel two   _____

☐ B channels
  ☐ Voice
  ☐ Data
  ☐ Both

ISDN Equipment Vendor Information

☐ ISDN Ordering Code (IOC) _____

ISDN WORKSHEET *continued*   *Page 3 of 3*

ISP Information
Configuration information

☐   Dial-up number: _____

☐   Login name: _____
☐   Login password: _____
☐   E-mail address: _____

☐   Assigned IP number: _____ . _____ . _____ . _____
☐   Gateway IP address: _____ . _____ . _____ . _____

☐   DNS IP address: _____ . _____ . _____ . _____
☐   Subnet mask: _____ . _____ . _____ . _____

☐   POP3 information
    ☐   Mail login: _____
    ☐   Mail password: _____
    ☐   Mail server: _____

☐   SMTP mail server   _____

Additional contacts: _____

_____

_____

_____

_____

_____

Notes: _____

_____

_____

_____

_____

To make the Worksheet easier to write on, use your copy machine's enlarged setting (for example, 129%).

What Do I Have Now?

Determining what you have involves checking ISDN availability in your area, your Internet access, and your COM port.

Local ISDN Availability

First, regardless of the widening availability of ISDN, you may still live in a neighborhood or geographic location where service is not available. This is easy to check: pick up the phone and ask. You can find the phone numbers and other ISDN carrier contact information in Appendix A.

If you already have an Internet connection using one of those "old" analog modems, you may want to check out your local RBOC's Web site (listed in Appendix A). Most RBOCs provide a form to look up availability in your area based on your current analog phone number.

Checking availability in your area may involve your local phone company running a loop qualification test. Be prepared. In some cases, this may be done overnight. In other cases, the company may need to schedule the test, and you will need to wait *two to three weeks* (or maybe even months) for the results. You will not know until you call.

After you establish availability, double-check the installation costs. As ISDN becomes more widely available, the pricing structures of the RBOCs have been changing frequently. Keep in mind that you will want to order the service plan that reflects how you will use ISDN, as discussed in Chapter 9.

ISDN Internet Access Availability

You also need to establish availability of an Internet service provider (ISP). First, determine if any ISPs are offering ISDN access in your area. If not, do not get discouraged—you may be able to find a national ISP. Be sure to determine if the ISP offers a local access number or *point of presence* (POP).

By choosing an ISP with a local POP, you can avoid long-distance toll charges. In some remote areas, you may not be able to get a local POP, and you will need to factor in the additional costs for long-distance access and online charges.

COM Port Availability

Chances are, if you were already using a modem, checking for an available COM port is a moot point. Also, you may want to keep your modem as a backup (you can connect it through the ISDN equipment's POTS port). That's right, we are assuming you will give up your modem to free a COM port.

But wait, is it a high-speed port? If you decide in the next part of the ISDN worksheet to use an external ISDN adapter (one that plugs into a PC's serial or parallel port), you will need to come back to this step and check the type of port your computer is using. One B channel or two, you will need to be sure that your old serial COM port can handle the increase in bit traffic. These ports impose certain limitations. Most PC serial ports will not transmit information faster than 115 kbps. This is less than ISDN's maximum 128-kbps speed. Also, the error-checking data transferred between the PC and your external adapter, required for analog data, results in overhead, which further slows speed to about 92 kbps. Finally, external ISDN equipment impacts your computer's CPU performance.

Is It a High-Speed COM Port?

Don't worry—this will be easy, just four quick steps:

1. Exit to the DOS prompt by exiting Windows.

2. Type **MSD** (for Microsoft System Diagnostics) at the prompt.

3. Click on COM Ports in the menu bar.

4. Check what type of UART chip is recognized under the COM port you intend to connect the ISDN equipment to.

Commonly Used Equipment (Authors' Choice)

In this section, we cover five ISDN terminal adapters that we have picked based on price, availability, bundled software, reputation, and support quality:

- 3Com Impact

- Supra NetCommander

- ISDN*tek CyberSpace Freedom Series

- Motorola's BitSURFR Pro

- US Robotics Sportster

After you browse through the information about these models and become familiar with some of the features, we will launch back into the second part of the ISDN online information access worksheet.

> **NOTE** Keep in mind that we have chosen five different vendors simply as examples, not to limit you to only the five following brands. The ISDN worksheet is designed to be generic and should apply to most ISDN terminal adapters.

3Com Impact

3Com Impact ISDN digital modems and adapters give users dial-up network access to the Internet (or the office). The external modem offers PC and Macintosh users simultaneous voice, fax, and data over a single ISDN line. Simply plug your phone or fax into the analog Phone Out port. There's even a version that includes a 14.4-kbps (V.32bis) analog modem for compatibility with non-ISDN services. The ISA adapter includes enhanced LAN access features and a bus interface for maximum throughput.

Figure 10.2 shows the 3Com Impact external modem.

FIGURE 10.2:

3Com Impact external ISDN digital modem (*photo courtesy of 3Com Corporation*)

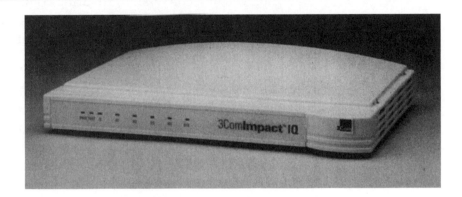

Product Information

The 3Com Impact product offers the following features:

- Fast dial-up access. Multilink PPP (MLPPP) combines both ISDN B channels into a single 128-kbps data connection for dial-up Internet or LAN access.

- Simplified installation. The installation program guides you step by step through the process, and there are no jumpers or switches. 3Com Impact's QuickSelect feature automatically decides which protocol to use for each call: V.120 for communications with other stand-alone digital modems, or Point-to-Point Protocol (PPP) for compatibility with ISP equipment.

- Optimized for your application. Choose the ISDN platform that meets your needs: the external modem connects to your computer's serial port and provides simultaneous voice, fax, and data. The ISA adapter provides a direct bus interface for maximum performance and ships with remote network client software and LAN drivers.

- Standards compliance. Certified for use in North America, the 3Com Impact digital modems support both V.120 protocol and Async-Sync PPP conversion, ensuring compatibility with ISPs and corporations. And the 3Com Impact ISA adapter ships with standard LAN drivers and supports a variety of international ISDN signaling protocols.

With MLPPP, 3Com Impact digital modems use dynamic bandwidth allocation to share bandwidth between your MLPPP session and phone or fax calls. In cooperation with 3Com's AccessBuilder 4000 with ISDN, your 3Com Impact ISDN adapter can automatically open a second B channel to handle a sudden transmission load, then transparently close the second connection when network requests are satisfied. Just as easily, an MLPPP session makes room for an outgoing voice call or fax transmission starting over the 3Com Impact external modem's Phone Out port, then resumes full bandwidth when the call or fax concludes. When you place a call or send a fax across the ISDN line, the bandwidth of the MLPPP session drops to 64 kbps. When the call is completed, the second B channel is restored, and the bandwidth of the MLPPP session returns to the top speed obtainable from your PC or Macintosh serial port.

NOTE

MLPPP is an Internet Engineering Task Force (IETF) standard designed specifically with ISDN and dial-up internetworking in mind. Like modem standards for compression or error correction, MLPPP must be used on both ends of a connection. The MLPPP standard is supported by many remote server vendors, ISPs, and commercial online services.

The digital modem is compatible with your favorite communications programs. 3Com Impact external digital modems support standard AT commands for easy and familiar operation. The modems come complete with a built-in NT1 termination device and everything else you need for quick connection to the ISDN network—including intuitive graphical configuration software, user manual, phone cable, and serial cable.

Ordering Information

When you order the ISDN line, give your carrier the configuration parameters for the 3Com Impact product you will be installing. You can do this simply by using the special ISDN ordering code (IOC) assigned to the product by Bellcore. IOCs are accepted by most phone companies, and they take the guesswork out of ISDN ordering.

TIP Recently, 3Com announced a program called IQ that guarantees Impact equipment installation in 15 minutes or less. If you do not have the equipment up and running in 15 minutes, you can return the equipment for a full refund of your purchase. There is one major catch to this service and guarantee: the ISDN line must be ordered from the 3Com service center. This is done to ensure that the ISDN line is available and configured accurately by the carrier.

An alternative is to send your carrier full parameter information. You can find tables of configuration parameters in the following:

- 3Com Impact documentation

- 3Com World Wide Web site: http://www.3com.com

- 3ComFactssm fax-back service: 408-727-7021

- 3Com BBS (set your analog modem to 8 data bits, no parity, and 1 stop bit): 408-980-8204

Call 1-800-NET-3COM within the United States and Canada to contact the manufacturer. (See Appendix B for other equipment vendor contact information.)

The 3Com Impact IOC is * J6 (for 3C861, 3C862, 3C871 and 3C876); the IOC for 3C872 and 3C877 is pending.

Supra NetCommander

The Supra NetCommander ISDN internal adapter is manufactured by Diamond Multimedia. This internal ISDN adapter combines good performance with some unique features at a reasonable price.

Product Information

Three software packages are included with the NetCommander adapter:

- Diamond AutoISDN configuration/monitoring utility
- Microsoft Internet Explorer
- Microsoft ISDN Accelerator Pack for Windows 95

Using Microsoft's Accelerator Pack, the NetCommander can reach an average speed of 112.9 kbps. The new Accelerator Pack version is incorporated with Diamond's software. With this combination, the adapter works with Windows 95's Dial Up Networking feature. Diamond helps with the task of installing newer drivers by posting the latest versions on the Internet (www.supra.com).

The NetCommander supports Windows 95 Plug-and-Play features to automatically set the interrupt (IRQ) and I/O base address to avoid conflicts. Windows 95's next maintenance release will support the installation of drivers when you boot your PC (currently, the "old-fashioned" method prevails, prompting you to install drivers from disks). The new drivers use Windows 95 Wizards to guide you through the setup process.

The internal adapter has the following features:

- BRI ISDN U interface (NT1 included)
- 56-kbps to 64-kbps data rates
- Voice and data on both B channels
- Caller ID enabled (where service available)
- MLPPP

The NetCommander uses only the bandwidth it needs. In low-traffic situations—for example, when you're reading a Web page—the NetCommander drops a B channel using just one 64-kbps channel. When you choose a file to download, the device opens both channels, doubling throughput to roughly 113 kbps. During business hours, when all RBOCs charge for connect time on each line, this feature translates into a big difference in your overall ISDN bill.

Diamond's Flex Channel Control lets you conduct analog phone (POTS) conversations while downloading files from the Internet. The analog interface uses a standard RJ-11 port. Finally, the Net-Commander has full ringing support, unlike other internal cards (which ring the computer or the card itself), so your regular phone can ring over an ISDN line. The adapter supports user-assigned distinctive ringing patterns.

Ordering Information

You can get information from the following sources:

- Supra contact information: http://www.supra.com/Support/contacts.html
- Supra fax-back service online: http://www.supra.com/Faxback/catalog.html

- Supra nontechnical information on products, services and other modem-oriented information: http://www.supra.com/Support/reference.html

- Supra file library of utilities: http://www.supra.com/Filel.html

- Supra product manuals: http://www.supra.com/Support/manuals.html

- Supra tech support newsletters: http://www.supra.com/Support/newsletter.html

The NetCommander requires IOC Capability Package M.

ISDN*tek CyberSpace Freedom Series

ISDN*tek offers the CyberSpace Freedom Series of ISDN access cards. Because these cards are designed exclusively for ISDN, they connect local calls virtually instantaneously, with tremendous speed and reliability advantages over analog modems.

Product Information

All the CyberSpace Freedom Series cards provide a WinISDN driver that can be addressed by application software through the industry standard WinISDN interface, for full utilization of all the bandwidth ISDN has to offer. Cards are available for Internet access, telecommuters, and for enterprise environments. An Integral NT1 version is also available for each type of card.

The CyberSpace Internet card has the following features:

- PC (ISA bus) hardware

- Windows drivers

- 56-kbps to 64-kbps data rates over one B channel; 128 kbps over two B channels using MLPPP software

- Compatible with Internet software and routers
- Versions for S or U interface ISDN connections
- Japanese INS64 (64 KB) version available

While Internet access and Web browsing are the most obvious applications for the Internet card, it also can be used for high-speed data transfers directly between two users, without using the Internet. If analog modems have been your typical method of file transfer, you will really appreciate the speed and reliability of ISDN peer-to-peer data transfers. To accomplish a peer-to-peer connection, both the sending and receiving computers must be ISDN-equipped. The software and hardware running on each machine should support an industry standard such as WinISDN or HDLC encapsulation using PPP protocols.

Here is the basic procedure for installing an ISDN*tek card if this is a new hardware installation:

1. Install the ISDNtek.sys miniport driver for Windows 95.

2. Install the ISDN Accelerator Pack software for Windows 95.

3. Once Dial-Up Networking is installed, open the Network icon from the Control Panel. Be sure that the TCP/IP protocol is installed.

4. Follow the ISDN*tek documentation to install the ISDN*tek adapter board and test the board with the ISDNTEST program.

Ordering Information

You can contact the manufacturer by regular mail, phone, fax, or e-mail:

P.O. Box 3000
San Gregorio, CA 94074
Telephone: 415-712-3000
Fax: 415-712-3003
e-mail: info@isdntek.com

ISDN*tek's Web page, at http://www.isdntek.com/Win95.htm, offers some good step-by-step instructions for installation. The ISDN*tek driver for Windows 95 is available on ftp.isdntek.com under the directory drivers/win_95/win95ras.zip or may be requested from info@isdntek.com with "Send Win95 Drivers" in the subject line.

Motorola BitSURFR Pro

Motorola's BitSURFR Pro ISDN modem is a high-speed terminal adapter. With two analog POTS ports, you can hook any two analog devices to the BitSURFR Pro: two telephones, a phone and a fax, a phone and an analog modem.

Product Information

To give you an idea of the procedures for setting up an external ISDN terminal adapter, the following sections provide the instructions for installing the BitSURFR Pro with Windows 95.

Installing the Configuration File

Before physically installing the BitSURFR Pro modem, use the following steps to copy the BitSURFR Pro configuration file, MOT1116A.INF, into the Windows INF directory:

1. Insert the Configuration Manager (version 1.3 or later) disk into the floppy disk drive.

2. From the Windows 95 desktop, open My Computer. Then open the floppy drive and the hard disk drive from the My Computer window.

3. From the hard disk drive window, open the Windows folder, select View from the pull-down menu bar, and choose Options.

4. In the Options dialog box, click on the View tab, then on the Show All Files option.

5. Click on the Apply button at the bottom of the Options dialog box, then on the OK button.

6. In the Windows window, double-click on the INF folder.

7. Make the floppy disk drive window active by clicking on it, and then select the file named MOT1116A.INF.

8. Grab and drag this file to the INF window. A dialog box will show that the file is being copied. When the copy is complete, close the INF window.

9. Make the Windows window active by clicking on it, and then select Options from View menu.

10. In the Options dialog box, click on the View tab, then on the Hide All Files of These Types option.

11. Click on the Apply button, then on OK in the Options dialog box.

12. Close the remaining open windows.

Installing the ISDN Modem

Use the following steps to install the BitSURFR Pro modem into a Plug-and-Play PC running Windows 95:

1. Shut down Windows 95 using the normal procedure and turn off your PC.

2. Plug the BitSURFR Pro modem data port into an available COM port.

3. Turn on your PC.

4. As Windows 95 boots, it will inform you that it has located new hardware and will identify it as a Motorola BitSURFR Pro.

5. After Windows 95 has booted, click on the Start button on the Taskbar.

6. Select the Settings entry from the menu, then click on Control Panel.

7. In the Control Panel, open the Modems folder and verify that the Motorola BitSURFR Pro is available.

These steps complete the installation under Windows 95. Your Windows 95 applications will now have the Motorola BitSURFR Pro as an available modem selection during setup.

Ordering Information

With SPIDs, TIDs, TSPIDs, DNs and ISPs flying at you from every direction, who do you call for help? The Motorola ISDN Life-GUARDs service is a no-charge ISDN help desk that guides you through:

- ISDN equipment and service selection
- ISDN line ordering and provisioning
- ISP subscription
- Equipment (ISDN modem/terminal adapter) installation

See the home page at http://www.mot.com/MIMS/ISG/Service_Support/lifeguards/

The modem is available in black (order code 6457504100010) and purple (order code 6457504100030) for the Windows version, and in black (order code 6457504100110) for the Macintosh version.

The BitSURFR Pro has been tested to operate with the following IOCs. Choose an IOC which is available from your ISDN service provider and best suits your needs. Be aware that using an IOC that provides capabilities you will not use may cost you more.

IOC	Voice	Simultaneous Voice & Data	Data up to 64 Kbps	Data up to 128 Kbps	Caller ID & Support Services
Motorola Access 3	Yes	Yes	Yes	Yes	Yes
Capability P	Yes	Yes	Yes	Yes	Yes
Capability S	Yes	Yes	Yes	Yes	No

Initializing Your Terminal Adapter

If you are familiar with using analog modems, you may remember choosing the modem's brand and model in your access software. When you install and configure your access software to use an ISDN modem, you need to select the make and model of your digital modem. If your modem is not listed, you may need to enter a *digital initialization*, or *AT command*, string for initializing your terminal adapter. The BitSURFR Pro ISDN modem digital initialization strings for V.120 and MLPPP applications (DOS/Windows or Macintosh) are listed below. Check the terminal adapter vendor's Web page or FTP site for up-to-date lists of ISDN terminal adapters and their initialization strings.

Application	At Command String
Single V.120 for DOS/Windows	AT&F&C1&D2\Q3%A2=2%A4=1
Single V.120 for the Macintosh	AT&F&C1&D0\Q3%A2=2%A4=1
MLPPP for DOS/Windows	AT&F&C1&D2\Q3%A2=95@B0=2%A4=1
MLPPP for the Macintosh	AT&F&C1&D0\Q3%A2=95@B0=2%A4=1

NOTE   Think of the initialization strings as a "secret code" (actually, if you are interested, you can decipher the codes by referring to the appendix found in the manufacturer's user guide) between your computer and the computer that will receive your call. The initialization string is sent to the remote or dial-up computer to tell it how you want to handle communications.

Windows 95 and the Get ISDN Program

Microsoft's Get ISDN Web site is designed to simplify and streamline the process of getting high-speed ISDN service for the Windows-based PC. This site is located at http://www.microsoft.com/windows/getisdn/.

Since this Web site opened in March 1996, it has received more than one million hits. In fact, the Get ISDN site is becoming a model for electronic ordering and is benefiting customers, local exchange carriers, and ISDN equipment vendors.

Customers benefit from the convenience of easy access to service provider information, faster order processing, and flexibility or ordering on a 24-hour, 7-day week basis. In fact, the RBOCs are reporting that approximately 75 percent of their orders are coming in after hours.

Local exchange carriers are able to dramatically reduce the average time a customer service representative spends on each order. Electronic ordering is a business's dream, because it offers an increase in order volume without an increase in labor costs. The carriers can provide faster and more accurate order implementation and installation, while expanding customer service hours around the clock.

ISDN equipment vendors or customer premises equipment manufacturers are enjoying an increase in sales with a reduction in customer support requirements.

This "win-win" situation is not going unnoticed. For example, the members of the NIUF (North American ISDN Users' Forum) want to expand the Get ISDN electronic-ordering concept and start a new, standardized process for ordering electronically. The NIUF's Mass Market Industries Group wants to eliminate the process of calling the phone company to order an ISDN line and allow people to order service from the Internet or even in-store kiosks. Imagine ordering ISDN at your local shopping mall! You can contact NIUF for more information at 301-975-2937.

US Robotics Sportster

US Robotics offers the Sportster ISDN terminal adapter, shown in Figure 10.3.

FIGURE 10.3:

US Robotics Sportster ISDN terminal adapter (*photo courtesy of US Robotics*)

Product Information

The Sportster terminal adapter has the following standards-based ISDN features:

- MLPPP to combine the 64-kbps B channels into a single 128-kbps channel. Dynamic voice override to automatically assign half the channel to a voice call, then return the full 128 kbps to data transmission when the voice call is done.

- Support for multiple compression types, including Stac, Microsoft, and Ascend. Software-based compression lets Sportster ISDN adjust to match the equipment on the other end of the call.

- TurboPPP for Windows 95 to provide the performance of MLPPP and compression (up to 512 kbps), even when the application supports only PPP (64 kbps).

- An integrated analog device port for a standard analog phone. Ringing occurs on the Sportster ISDN board. (An optional external ring generator cable is required for fax machines or modems to answer incoming calls.)

- Simultaneous voice and data to allow you to talk on the phone or send/receive faxes while transmitting data at 64 kbps. A special ISDN phone is not required.

- An integrated NT1 to allow direct connection to the ISDN service provider's wall jack, without a stand-alone network termination device. This saves the space and cost of an external NT1.

- Multiple interfaces, including NDIS, ODI (for IPX), Packet Driver, WinISDN (for Internet access), Windows 95, and TAPI (Telephony Applications Programming Interface) support. These let you use Sportster with off-the-shelf Internet, remote LAN access, and data communication software packages.

Ordering Information

For pricing and availability information, call the US Robotics Corporate/Systems Sales department at 1-800-USR-CORP. For ISDN line ordering information and technical support, call 1-888-USR-ISDN or send Internet e-mail to support@usr.com.

Who Supplies the Equipment?

At this point, you should have determined ISDN availability in your area and, ideally, found several ISPs providing ISDN dial-up. You should also have familiarized yourself with the hardware offered by some vendors.

Now we've arrived at the second part of the ISDN worksheet. We will examine equipment issues and features. Our goal is to keep you on track for a safe landing—a successful, compatible ISDN setup.

What the ISP Provides

Will your ISP provide equipment? As a service to customers, some ISPs are offering to help customers provision their equipment. In this situation, you will only need to shop for a competent ISP.

Equipment Recommendations

Because you and your ISP will be connecting to each other, we suggest that you ask their representatives about the equipment they have had the most success with for connecting to the Internet. A competent and experienced ISP will have connected with a number of users with various types of equipment. Your ISP should be able to provide valuable information to help your decision-making process.

If you are following the steps outlined in this book, you will know important details about your current computer configuration. This information and the recommendation from your ISP will help you choose the most compatible customer premises equipment and the most user-friendly for your type of operating system.

ISP Shopping Parameters

Initially, shopping for an ISP involves three basic parameters:

- The basic monthly charge for the service
- The usage included in the basic charge
- The charges for additional usage

Typically, an ISP will charge a fee per month. This fee includes a set allotment of hours or megabytes. In some cases, usage may vary. Normally, a provider will charge by the hour, or more precisely, by the B-channel minute. Often this can be confusing. An ISP may include one or two B-channel access in the basic monthly charge. Other ISPs might offer one B channel with the option of adding another B channel, if available. In other words, if the number of ISDN users and the corresponding B channels used are low, you can add another ISDN B channel.

For example, suppose that you sign up for a service that offers 50 hours of usage per month included in your basic monthly charge. During the month, you surf through some bandwidth-intensive sites, and your terminal adapter is the type that adjusts so that those connections use two B channels. At the end of the month, you find additional charges for exceeding your monthly allocation. Instead of 50 hours of online time, you "burned" your monthly allocation twice as fast using two B channels and enjoyed only 25 hours of online time.

Be sure to confirm how the ISP plans to bill you for B-channel minutes. In some cases, ISPs may choose to round to the next minute rather than charging for a partial minute. This can add up if you plan to configure your ISDN terminal adapter to disconnect when your line is idle and then reconnect when you decide to move on. As always, the old adage applies: buyer beware.

In a few cases, you may find that the ISP charges for the number of megabytes used. This is rare.

Also, you may find ISDN Internet access that provides unlimited access for a flat or fixed monthly rate. But what may look like a good deal at first may turn out to be frustrating later. Other users flock to take advantage of the provider's attractive package. Very soon, you may find it difficult or impossible to access the provider because the growth and popularity of that service results in many people dialing into the ISP server. Because of the flat rate, they have little or no incentive to disconnect to make their port available to the other users in your ISP's community.

A low basic monthly charge may have higher charges for additional usage. Make sure, and record the information in the ISDN worksheet.

Bundled Services

Another item to check out with the ISP is bundled services. In addition to ISDN Internet access, some ISPs offer bundled services, or other features that come along with your dial-up package. These features may include Web page storage, an FTP drop box, Domain Name Service (DNS) support, and other attractive options. Ask each ISP what other services are provided.

> **NOTE**
>
> Do you want to connect to your company? Are you wondering about dialing-up your company to download files or access the company's server? Are you planning on telecommuting (working from home)? If you answer "Yes," or even "Maybe," to these questions, turn to Chapter 11. Your needs extend beyond the scope of this chapter, which deals with your basic ISDN dial-up access to the Internet.

What You Provide

We've suggested some commonly used equipment and explained that it's a good idea to ask ISPs for their recommendations. Here, we want to stress that you should be sure to determine your needs in two areas: compression and MLPPP+.

Compression

Suppose that you decide that you want or need compression. The numbers look impressive: 2:1 compression delivers 512 kbps to your door. Be careful for two reasons. First a salesman's enthusiasm may fail to inform you that most of the information you will encounter on the World Wide Web is already compressed. Images on Web pages and video clips from your favorite movie are often compressed. As you might suspect, you cannot compress media that is already compressed. So compression will not offer much of a performance advantage to the Web surfer. Second, if you believe you need compression, then maybe you are considering connecting to your company's LAN or telecommuting. Chapter 11 is the place for you. Turn to the worksheet in that chapter for more information.

MLPPP+

Do you want MLPPP+? As discussed earlier, this is a protocol that many ISDN equipment vendors are adopting. MLPPP+ allows your ISDN terminal adapter to dynamically add the second B channel based on user-defined parameters. The availability of this feature is increasing, but currently it is proprietary and supported by only a few vendors. Be sure to share your wishes with your ISP representative; he or she will tell you if the ISP's equipment offers the feature or not.

What Information Do I Need from My Provider?

The last sections of the online information access worksheet cover the information you should receive from your local exchange carrier, equipment vendor, and ISP.

Local Exchange Carrier Information

The contact information, installation date, and service order number you receive from your local exchange carrier are items that usually get lost in the shuffle of implementing ISDN. It also helps to work with the same person to resolve any problems quickly.

You also should confirm rates and usage charges. In addition to monthly charges, the usage rates will vary by the time of day. Although you will be making local calls, the tariff structure for ISDN usually includes per-minute rates during certain times, unlike most residential analog service, which includes local calls within a certain zone.

The Installation Information section of the ISDN worksheet may look like a repeat of the Contact Information section, but often the person who takes the order is not the person managing the installation. When applicable, record the person or persons that will be doing the installation at your home or business. This installation information will be useful if you experience any problems.

Provisioning Information

At this point, you may be wondering if all of these details are really necessary. We can understand your impatience and your desire to get ISDN up and running. We expect that some of you will even skip some of the previous sections of the ISDN worksheet. Well, here is your chance to avoid time-consuming problems and frustration when you are ready to connect your new terminal adapter to your new ISDN line.

We cannot stress the importance of the provisioning data. If things go wrong, often the problem can be traced to incorrect provisioning information. Why? Because your provisioning information answers the questions that the configuration software needs to enable your ISDN terminal adapter to synchronize (show green lights on the panel, if it is external) with the switch.

First, ask the customer service representative who is handling your installation for the following information. Second, be sure to confirm this data with the installer when he or she comes to activate your ISDN line. This is a key step. The installer can confirm your actual situation, which can easily change between the time of your order or request and when the service is actually installed.

- **ISDN type:** As discussed in Chapters 3 and 9, your ISDN may be either National or Custom.

- **Digital switch:** Record the ISDN switch type. See Chapters 3 and 9 if you want to review the information about the various switches used by phone companies.

- **Local directory numbers:** These numbers are your phone numbers. An ISDN line can have more than one directory number.

- **SPIDs:** Here you need to put your most assertive self forward and push for *formatted* SPIDs. Your local exchange carrier has the capability to look at the switch you will be connecting through and see exactly how the SPID is formatted (that is, how many ones or zeros are placed in front and in back of the local directory numbers). See Chapter 14 for more information about formatted SPIDs.

- **B channels:** Record how your B channels are configured. If the phone company does not charge extra, specify both: voice and data. This will simplify the configuration process that comes later.

Equipment Vendor Information

The information you need from your equipment vendor is the IOC for your ISDN terminal adapter. Having this number will help your phone company and you get the configuration right. See Appendix B for equipment vendor contact information.

NOTE Some of you are already saying that keeping track of SPIDs, switch types, and ISDN types is lots of trouble. But reserve some compassion for your local exchange carrier workers. Often, they will need to set more than 25 parameters when provisioning the switch for your "single" ISDN connection.

ISP Information

In many cases, ISDN customers forget that there are two connections involved in high-speed Internet access. The connection that gets the most focus is the one to your ISP. The other, often overlooked connection is the one to reach your ISP from your home or office. After you are finished installing and configuring your ISDN terminal adapter, you need to configure your Internet access software. You should obtain the following information:

- **Dial-up or access phone number:** The number your ISDN access software will dial to connect to the Internet.

- **Login name:** What you enter to log in to your PPP or MLPPP account. Typically, the name can be up to eight characters. Keep in mind that CaSe counts.

- **Login password:** Your password is entered after the login name. Again, typically this is eight characters or less and case matters.

- **E-mail address:** How others can send e-mail to you. Often, the login name is the same as the name listed before the @ sign. E-mail addresses are not case-sensitive.

- **Assigned IP number:** The IP (Internet Protocol) address for your computer. This is assigned by the ISP. In fact, when you connect to the Internet, your computer will increase the Internet by a size of one while you are connected. Your IP address is your node on the worldwide Internet.

> **NOTE** In increasing numbers, ISPs are using dynamic IP number allocation. If this is your ISP's method, you will not receive an actual IP number, but you will need to find the appropriate button or checkbox in your access software's configuration to set up this method.

- **Gateway IP address:** The machine or server at the ISP location that will route traffic to and from your computer to the Internet.

- **DNS (Domain Name Server) IP address:** This is the database server located at the ISP that handles Domain Name Service retrieval functions. Queries from your computer to this machine convert domain names, such as SYBEX.COM to numeric IP addresses. Typically, the DNS IP address will be in numeric format.

- **Subnet mask:** Usually, this receives a default value of 255.255.255.0 for single dial-up users. In some situations, covered in Chapter 11, you may be setting up an account for a small group of machines or LAN. The subnet mask will tell your ISP how to properly route data to the computers on your LAN.

- **POP3 information:** This tells your computer how to manage your e-mail information. The mail login may or may not be different than your login name. Again, uppercase and lowercase matters. The mail password will be used to access your POP3 e-mail account (yep, you guessed it, case matters). Your POP3 mail server is the name of the computer that stores your incoming e-mail.

- **SMTP mail server:** This is the name of the ISP's machine that forwards the mail you send, or outgoing e-mail. It may or may not be the same machine as your POP3 mail server.

After you've completed the steps in your ISDN worksheet for your online information access application, you're ready to sit back and enjoy the results of your work. We're sure you'll appreciate your ISDN connection as you zip around the Web.

A Guide to ISDN Setup for Telecommuting and RLA

- LAN security and access requirements

- RLA component optimization

- Tariff management techniques

- Selected high-performance LAN, Internet access, telecommuting/mobile worker, and interoffice connectivity solutions

- Information from your service providers

With the increase in the number of telecommuters and company employees working from remote areas, it is becoming more likely that you will be working for a company that requires remote connectivity, with or without a LAN connection.

We've combined the guide for ISDN setup for telecommuting and remote LAN access (RLA) because the equipment and general procedures for these two applications overlap. The same equipment vendors, and in some cases, the same equipment, can satisfy both telecommuting and RLA needs.

Again, the only difference between the two applications is that the RLA user needs to be concerned about LAN interface and protocol issues. The telecommuter may or may not be connecting to a company or corporate LAN.

Which Application Do I Want?

This chapter looks at two applications: RLA and telecommuting. For purposes of our discussion, you may want to think about the two applications as follows:

- Telecommuting is primarily a workstation-to-central company connection (one to many).

- RLA is primarily a workstation-to-LAN or LAN-to-LAN connection (many to many).

Telecommuting

If your telecommuting involves connecting to a LAN, we suggest you refer to the online services worksheet in Chapter 10 (Figure 10.1) as well as the one in this chapter for guidance in connecting your small office home office (SOHO) to your company's or client's LAN. You may not realize it, but dialing up to connect to your ISP is similar to making a connection to a company's LAN.

To keep it simple, you can follow the steps prepared for Chapter 10 with only two differences:

- You will want to use the equipment recommended or supported by the company's network administrator. Because your network administrator must support equipment attached to the LAN, he or she will appreciate your cooperation (or demand it) in having a system that is compatible with others that are connecting to the LAN. Equipment requirements are discussed in this chapter.

- In the online service access worksheet, substitute the name of the LAN for every reference to an ISP. Remember, in most cases, your telecommuting connection to your company's LAN will be very similar to a connection to an ISP's LAN.

WARNING

Make sure that your analog equipment (answering machine, phone, and fax) is compatible with your analog conversion jack(s) on your ISDN terminal adapter equipment.

RLA

Unfortunately, there is no single "best" solution or path for implementing an RLA plan, because there are so many options and variables involved in remote access to a LAN. Our ISDN worksheet for RLA applications is more like a checklist than a recommended series of sequential steps. This worksheet is presented in Figure 11.1.

F I G U R E 11.1 : ISDN RLA worksheet

ISDN WORKSHEET	Page 1 of 4

Which Application Do I Want ?

    Remote LAN Access

What Do I Have Now?

Local ISDN Availability

☐ Is ISDN available from local phone company?

☐ Is a loop qualification needed?
    Completion date ___/___/___

☐ Confirm installation charge $_____

☐ Does the local phone company offer a special package for RLA?
☐   If so, what packages and vendors are involved?

Destination ISDN Availability

☐ Is ISDN available at destination sites?

What interfaces do you need to support?
☐   ODI
☐   NDIS
☐   Other_____(verify with equipment vendor)

What LAN protocols do you need to connect to?
☐   TCP/IP
☐   IPX
☐   Ethernet
☐   Appletalk

What security methods are used?
☐   None
☐   Security card
☐   Dial in
☐   Dial back
☐   Restricted resources, e.g., databases, applications or users (access denied or limited)?
☐   Special security software

What are the login processes?
☐   None
☐   Single, standard login process
☐   Login for specific applications
☐   Login for network connection
☐   Special security-related login process (verify with network administrator)

ISDN WORKSHEET *continued*

☐ Are there access limitations?
   ☐ Special hours of access?
      If so, from_____to_____
   ☐ Time limit on connection?

☐ Is there support available?
   ☐ Contact information for support
      Name: _____
      Phone: _____
      E-mail: _____
   ☐ Times and days support is available? _____

☐ Review organization's RLA policies with network administrator

RLA Optimization Areas

☐ Remote computer
   ☐ PC serial port
   ☐ Network applications
   ☐ Operating system

☐ Network operating system

☐ Remote access client software

☐ Remote links

☐ Remote access security

☐ Remote access server
   ☐ Hardware
   ☐ Software

☐ Network management

☐ LAN efficiency

Tariff Management

☐ Bandwidth control
   ☐ Bandwidth on demand
   ☐ Bandwidth aggregation and augmentation
   ☐ Switchover
   ☐ Minimum call duration timer

☐ Connection control
   ☐ Time-of-day tariffs
   ☐ Callback
   ☐ Prioritization

☐ Data control
   ☐ Data compression
   ☐ Triggered routing protocol updates
   ☐ Spoofing

ISDN WORKSHEET *continued*

Who Supplies the Equipment?

What Your Company Provides

☐ See your network administrator

☐ Equipment provided?

_____
_____
_____

☐ ISDN service provided?
  ☐ Installation
  ☐ Usage charges

What You Provide

☐ See your network administrator

☐ Equipment required or recommended

_____
_____
_____

☐ Vendor or retail outlet with discount plan for your company?

☐ Local carrier package approved by your company?

☐ ISDN service
  ☐ Installation
  ☐ Usage charges

What Information Do I Need from My Provider?

Local Exchange Carrier

☐ Contact information
  Phone company rep: _____
  Company name: _____
  Phone: _____

☐ Installation date: _____
☐ Service order number: _____

☐ ISDN service charges
☐ Monthly charge $_____
☐ Per minute usage
    8 a.m. to 5 p.m. $_____
    from ____ to ____ $_____
☐ Additional charges $_____

☐ Installation information
  Phone company rep: _____
  Company name: _____
  Phone: _____

  Provisioning information
☐ Type of ISDN
  ☐ National
  ☐ Custom

ISDN WORKSHEET continued

☐ Digital switch type
  ☐ 5ESS
  ☐ DMS-100
  ☐ Siemens EWSD

☐ Local directory numbers
  ☐ B channel one _____
  ☐ B channel two _____

☐ Service profile identifiers (SPIDs)
  ☐ B channel one _____
  ☐ B channel two _____

☐ B channels
  ☐ Voice
  ☐ Data
  ☐ Both

ISDN Equipment Vendor Information

☐ ISDN Ordering Code (IOC) _____

Information from the LAN Administrator

Configuration information

☐ Dial-up number: _____

☐ Login name: _____
☐ Login password: _____
☐ E-mail address: _____

☐ Assigned IP number: ____.____.____.____
☐ Gateway IP address: ____.____.____.____

☐ DNS IP address: ____.____.____.____
☐ Subnet mask: ____.____.____.____

☐ POP3 information
  ☐ Mail login: _____
  ☐ Mail password: _____
  ☐ Mail server: _____

☐ SMTP mail server _____

Additional contacts: _____

_____

Notes: _____

_____

As we will explain in this chapter, you need to consider many factors during the design of your RLA system. The ones that are most important for your situation depend on your organization's unique needs and priorities. The worksheet in this chapter should give you a good idea of the many contingencies involved, as well as your options.

What Do I Have Now?

Because an RLA system may be connecting several branch offices, telecommuters, and ISPs, you will want to be thorough in confirming ISDN availability for all sites. Your big advantage is the economy of scale. Connecting several sites and many users creates a strong position for you or your company to negotiate an attractive package or reduced rates. Be sure to conduct a careful inventory of who and what you have before approaching the task of building an RLA system.

Local ISDN Availability

Along with verifying that ISDN is indeed available in your area, you should discover if the local phone company offers a specially priced package for remote LAN access or telecommuting. If the answer is yes, your next question is, "What are the packages and vendors involved?"

Destination Access Availability

For telecommuting and RLA applications, you must also check the availability of ISDN in your destination area or areas. See Appendix A for ISDN carrier contact information.

LAN Interfaces and Protocols

You'll need to determine LAN-related information for your destination access. These items include the interfaces you need to support and the LAN protocols that you need to connect to.

Common interfaces are ODI (Open Datalink Interface) and NDIS (Network Driver Interface Specification). Common LAN protocols are TCP/IP (the suite of protocols developed for use over the Internet), IPX (the protocol used by Novell NetWare), Ethernet (for Ethernet networks), and Appletalk (for Macintosh networks).

LAN Security

LAN access issues include security methods or security access policies. You'll need to know the corporate policy for security. Find out if you need a special security card (this type of card looks a lot like the ATM card that you use at the bank) and software. You also need to know if you dial in or if it is a dial-back connection.

Your company policy may be to restrict some resources, such as databases, applications, or connections to other uses, by limiting access to them or not allowing access to them at all. You'll want to be sure which resources will be available.

Login Processes

Another consideration for RLA applications is login procedures. There may be one standard login process, or separate login procedures for specific applications and your network connection. You may also need to go through a special login procedure that is tied to your company's security policy.

Access Times

Some companies will not let you connect to their LAN whenever you feel like it. They have designated special hours of remote access. You may also find that there is a time limit on each of your network connections. These restrictions may be due to LAN traffic considerations or part of a company's security policy.

Corporate Policy

Be sure to review any company or corporate policy for remote access or telecommuting. Some policies include provisions that affect your ISDN service or ISDN equipment configuration. Talk to your network administrator!

RLA or Telecommuting Support

For your RLA or telecommuting application, support can be very important. Find out what support is available to you. Is vendor or integrator support available? How do you contact the support service? Is there full-time support available, 7 days a week and 24 hours a day, or are there special hours?

Optimizing RLA System Components

The cliché that a chain is only as strong as its weakest link applies to the world of RLA. Between the organization and the remote user, road warrior, or branch office lies a complex combination of PC hardware and software, internetworking, and telephony. Therefore, to implement an efficient system, each component in an RLA system must be addressed.

In the following sections, we will look at the optimization of several key components, each of which play a role in the integration and delivery of information over a network.

Remote Computer

A key consideration for a remote computer is that it must function as a stand-alone system when it is not connected to a network. The remote computer must perform adequately—have enough CPU power, memory, and disk storage—to handle the end-user's applications.

PC Serial Port   A significant but often overlooked component of the remote computer is the serial port. Until recently, the serial port was limited in speed to 9600 bps and below. Improved technology has pushed the new point to 115,200 bps. This is an improvement, but it's still short of the 28,000 bps that ISDN delivers. You may need to factor the serial port into your overall design as a possible bottleneck.

Network Applications   Network applications may impact performance more than any other component of an RLA system. A network application that is not optimized for remote access may be designed to look at the network server as a file system. This "network-as-file-system" model ignores issues such as bandwidth and costs of moving lots of data across a remote access system. This approach to network applications assumes you will only communicate or exchange data over a LAN. It was not designed for connecting a LAN to other computers or LANs across the country.

Consider network applications that are built on the client/server model. Because client/server applications send only the data that the client (user) needs over the network, they use bandwidth efficiently and reduce data-transfer costs. Other options to consider are remote-control systems that can be layered on top of a standard network-as-file system.

Remote Computer Operating System   The remote computer's operating system affects the maintenance and establishment of compatible connections while preserving end-user usability and convenience. This option needs to balanced with the network operating software requirements and limitations, discussed in the next section. For example, will the network operating system allow the client programs to load and stay resident, without needing to open a connection, or to load upon demand.

NOTE

A program that "loads and stays resident" is one that is loaded into the local computer's memory and stays there, quietly, not executing, until an event or command from you brings it to life. You could think of it like a bear hibernating, until you disturb it from its sleep.

Network Operating System

Again, the interdependency among components becomes evident during an examination of the protocol implementation. The speed of a remote access system is a function of the system's bandwidth and the ability of the system to use the bandwidth efficiently. An ideal is to keep the "pipe" filled with essential information. The protocol, or the way the computers will communicate, must be compatible with your network operating system. Be sure to choose one that supports a wide variety of protocols to maintain flexibility for your RLA program.

Remote Access Client Software

The remote access client software is the user's main link into the remote access system and affects the remote access system's ease of use. In addition to providing a user-friendly environment, the client software must be compatible with other components in the remote LAN environment. This software also has an impact on the system's overall speed. Be sure that your remote access client is compatible with multiple protocols and supports them simultaneously.

Remote Links

As mentioned earlier, networking applications are designed to work at a LAN's high bandwidth. Logically, the faster the remote link, the less the difference between the local and remote machine's performance. As you know, ISDN provides a much faster connection than traditional analog modems. In addition, ISDN cost benefits really shine in a dial-up business case. This is especially true for remote users who dial in from fixed locations (full-time telecommuters). An optimized remote access system should take advantage of digital connections whenever possible.

Remote Access Security

Unfortunately, there is no single "best" or optimal solution for remote access security; a large number of factors must be considered. Some of the trade-offs include:

- Optimal security system installations should be simple, with minimal setup time.

- The server's internal database should store user names and passwords, as well as attributes for each user.

- As a rule-of-thumb, in larger-scale systems, trade lengthier initial setups for long-term time savings in system management.

- A remote access server should provide a simple and seamless integration of external security systems. This integration will increase your flexibility in implementing security options and security schemes.

Remote Access Server

The following considerations should be made for hardware used as a remote access server:

- Server hardware should be optimized for remote access. Avoid a general-purpose hardware platform.

- Server hardware should be optimized for serial port throughput and CPU power.

- Focus on integration to increase convenience, improve support, and simplify ongoing maintenance.

Like your client software, the remote access server software should support a wide variety of protocols and addressing schemes. The remote access server software impacts the performance and throughput of the entire system. Minimizing overhead and reducing unnecessary broadcast traffic are procedures that are specific to each protocol and must be considered to optimize your system.

Network Management

Network management for a remote access system should be optimized to minimize the network manager's involvement in setup and maintenance. Which factors to consider will vary based on the size of the installation. For example, in a small system, speed, simplicity, and interactive use are important to consider. In a large system, power, automation, and integration are significant.

LAN Efficiency

Finally, remote access systems require that the network manager monitor and gauge the efficiency of the LAN. In general, internetworks that are routed perform more efficiently than bridged networks. A LAN segment that carries a significant amount of traffic may need to be further segmented via routing technology to provide a higher level of performance, both on a local and remote level.

Tariff Management

Tariff management involves the management of bandwidth, connections, and data control. Applying all three techniques intelligently will help the network manager to contain and control the costs of the RLA system.

Bandwidth Control

By deploying dynamic bandwidth-on-demand techniques, bandwidth control maximizes network efficiency and minimizes cost. There are four key areas of bandwidth control:

- **Bandwidth on demand:** A connection is opened only when there is data to send, and then it is closed immediately afterward. Choose equipment that provides timeout parameters (these parameters are usually found in the configuration software). The values you set will depend on your local exchange carrier's tariff and the applications that your network will be using.

- **Bandwidth aggregation and augmentation:** Extra data channels (B channels) are added when they are needed, and shut down when they are not. On-demand bandwidth using ISDN B channels is commonly found in applications that augment leased lines or accommodate overflow traffic. In addition, you may wish to consider configuring lines for optimizing throughput by setting parameters to determine when a line should be opened and when it should be closed. Often, it is wise to build in some latency (delay or wait time), to accommodate "bursty" traffic.

- **Switchover:** This feature determines when the network traffic moves from one circuit to another. When traffic begins to saturate the bandwidth of a leased line, traffic can be "switched over" or moved to ISDN. Later, when network traffic returns to normal, the ISDN line is shut down. This is definitely a feature to consider when you want to be sure that network traffic is traveling on the most cost-effective circuit.

- **Minimum call-duration timer:** This is an extension of the bandwidth-on-demand timeout feature. For example, most call tariffs may have a minimum call time or an initial-minute charge. In either case, you will want to configure how calls are handled to minimize costs.

Connection Control

By prioritizing connections and minimizing costs under different tariffs, connection control optimizes connections to remote locations. Connection control also provides fast recovery from failure. There are three key areas of connection control:

- **Time-of-day tariffs:** An obvious but often overlooked consideration. Different tariffs are charged based on the different times of the day: higher during the day and lower during the evening. A network manager may choose to make use of ISDN at night for data replication and backup, while minimizing use during the daytime hours.

- **Callback:** This is another tariff-based technique, which is employed when the tariffs between two remote locations vary. This technique is often used in managing costs for international calls. ISDN offers several advantages when using this technique. First, CLI (calling line identification) can be used for the callback number. Because the initial call can be refused, no charge is incurred and the CLI is used to return the call. Also, PPP-based PAP (Password Authentication Protocol) or CHAP (Challenge Handshake Authentication Protocol) can be used to identify the remote site to be called back.

- **Prioritization:** Circuits can be prioritized. For example, key remote locations can be guaranteed bandwidth at the expense of less critical ones.

Data Control

Using techniques such as spoofing and triggered routing protocol updates, data control provides the most efficient use of available bandwidth. For example, it manages whether usage-sensitive lines, such as ISDN connections, are not left open when no data is being sent. Also, data control uses data compression to squeeze the most data possible through the most available pipe. There are three key areas of data control:

- **Data compression:** Unfortunately, due to the complexity of considerations in applying data compression to a network, no optimal solution exists. The following are several reasons to consider compression:

    - To continue using lower-speed lines despite increasing bandwidth requirements. For example, averaging 2:1 compression transforms a 64-kbps line into a 128-kbps line.

    - To improve latency across a low-speed line—less data, less latency.

    - To reduce network costs on a usage-sensitive service such as ISDN.

- **Triggered Routing Protocol:** Routers and servers communicate constantly with each other. Routers broadcast their routing tables. Servers broadcast available services. Costs can soar if these broadcasts are allowed to travel freely. Stop the broadcasts, and the routers and servers cannot communicate. Fortunately, several methods exists to minimize unnecessary broadcasts. But beware of flawed techniques, such as piggybacking and timed updates, which are not really effective for data control.

- **Spoofing:** Like BONDING, spoofing is another term that is often used incorrectly. *Spoofing* refers to a set of techniques to keep service packets or network housekeeping information off the WAN link, while simultaneously fooling the network into believing that the messages or frames are being sent. Consider how your prospective internetworking vendors handle "keep-alive" packets. This is another feature that will minimize costs when channeling data across a usage-sensitive service such as ISDN.

Commonly Used Equipment (Authors' Choice)

Your equipment selection depends on what primary function it will serve. Will your main work involve dialing up to download files, accessing the company network server, or telecommuting (working from home)?

RLA products offered by three equipment vendors are described here:

- Ascend Communications
- Shiva Corporation
- Cisco Systems

We'll take a look at what these companies offer for high-performance LAN, Internet access, interoffice connectivity, and mobile/telecommuting solutions.

Ascend Communication Products

The Ascend Communications products for RLA and telecommunications applications described here include the Pipeline and Max product lines.

High-Performance LAN: Pipeline 400 Remote Access Server

The Ascend Pipeline 400 remote access and Internet access servers allow you to connect a corporate enterprise network to remote sites and telecommuters. The Pipeline 400 T1/E1 remote access server supports up to 30 concurrent analog or digital dial-in users. The Pipeline 400 BRI remote access server provides four ISDN interfaces, supporting up to eight concurrent analog or digital dial-in users.

Pipeline 400 products support IP and IPX routing and bridging, inverse multiplexing, data compression, and dynamic bandwidth allocation. They include LAN and WAN network management tools, such as MIB I and II support, Telnet remote management, call detail reporting, usage capping, WAN loopbacks, and flash memory.

The servers have a RISC-based architecture that can support PPP, SLIP, X25, and MP+. They also support security protocols that allow password encryption, restricted access based on IP address, and user privilege levels.

Internet Access and Mobile/Telecommuting Solutions: Ascend Pipeline 25–130

The Ascend Pipeline product line provides remote access to home office users, telecommuters, mobile workers, and remote offices. Table 11.1 summarizes the main features of these Pipeline products.

TABLE 11.1: Ascend Pipeline Product Features

Feature	Pipeline 25-Fx	Pipeline 25-Px	Pipeline 50	Pipeline 75	Pipeline 130
Bridging	Yes	No	Yes	Yes	Yes
IP routing	Optional	Yes	Yes	Yes	Yes
IPX routing	Optional	No	Yes	Yes	Yes
PAP & CHAP security	Yes	Yes	Yes	Yes	Yes
POTS	2 RJ-11	2 RJ-11	No	2 RJ-11	No
S/T interface	Yes	Yes	Yes	Q3	Q3
U interface	Yes	Yes	Yes	Yes	Yes
Number of users	Up to 4	1	Unlimited	Unlimited	Unlimited
PPP	Yes	Yes	Yes	Yes	Yes
MLPPP (RFC 1717)	Yes	Yes	Yes	Yes	Yes

TABLE 11.1:   Ascend Pipeline Product Features (continued)

Feature	Pipeline 25-Fx	Pipeline 25-Px	Pipeline 50	Pipeline 75	Pipeline 130
MP+	Yes	Yes	Yes	Yes	Yes
SNMP	No	No	Yes	Yes	Yes
Data compression	Optional	Optional	Yes	Yes	Yes

Interoffice Connectivity: Ascend Max 200Plus and 4000

The Max 200Plus is an eight-port WAN access switch that uses industry-standard PCMCIA card technology and integrates support for both analog modems and ISDN. This product offers a solution for providing e-mail, Internet access, and file-transfer capabilities to small- and medium-sized businesses, mobile workers, and telecommuters.

The Max 200Plus provides support for IP and IPX routing, and bridging for all other LAN protocols. It include standards-based security, network management tools, and dial-out modem and fax capabilities.

> **TIP**   Off-the-shelf, self-configuring PCMCIA modem cards for the Max 200Plus are available from third-party vendors.

The Ascend Max 4000 is a high-performance WAN access switch that lets corporations extend their corporate backbone networks to sites such as remote offices, telecommuter locations, and other corporations. The Max 4000 uses a single set of digital-access lines to provide 96 simultaneous dial-up connections, 48 of which can originate from analog modem users.

It supports standard protocols, including TCP/IP and IPX, PPP, and SLIP, and bridges all protocols (BCP standard bridging). The

Max 4000 also has bandwidth-on-demand features, allowing bandwidth to be controlled manually, automatically, or by time-of-day profile.

Ordering Information

Here is the contact information for learning more about Ascend's Pipeline and Max products:

World Wide Web site: http://www.ascend.com

BBS: 408-980-8204 (set your analog modem to 8 data bits, no parity, and 1 stop bit)

Shiva Corporation Products

The Shiva Corporation products covered here are the LanRover Access Switch, WebRover, Integrator 200, and AccessPort.

High-Performance LAN: LanRover Remote Access Server

The LanRover remote access server has a multiprocessing switching architecture and provides high-bandwidth WAN access. It has 128 KB EPROM; 2 MB battery backed-up SRAM; and thick (10Base5), thin (10Base2), and twisted-pair (10BaseT) Ethernet connectors.

The LanRover server supports IPX dial in with NetWare VLMs and NETX (includes all necessary software), as well as AppleTalk/PPP dial in.

Internet Access: WebRover

The WebRover Stack is a solution for remote, Internet. and intranet access. It allows remote employees to access information on the company LAN, private Web server, and the Internet by dialing one number. WebRover also provides the corporate site with dedicated, high-speed, leased-line Internet connectivity, for both LAN-based and remote workers.

Corporate sites can choose a WebRover Stack with 8, 16, or 24 analog ports, or a 20-port configuration that supports a mix of analog and ISDN.

Interoffice Connectivity: Shiva Integrator Routers

The Shiva Integrator 100 router is suitable for small or branch offices. It supports BRI and provides two channels for connection to other branch offices or to the central network with up to 128 kbps of bandwidth.

The Shiva Integrator 200 router routes both IP and IPX. It allows connections to and from more than one IP network over a single interface, allowing you to merge two distinct networks. For example, users can access both the central LAN and the Internet. While the remote access server provides multiple security options for dial in, the Shiva Integrator 200 router adds IP filtering at the packet level to allow or prevent access based on IP addresses. In addition, the firewall software prevents public users from gaining entry into the corporate LAN through application-level filtering. You can use the Shiva Integrator 200 in environments such as Windows NT, where addresses are dynamically allocated with DHCP from a remote server.

The Shiva Integrator 500 router is a powerful ISDN concentrator for central sites. It supports PRI, providing up to 46 simultaneous connections to remote locations. It also incorporates a V.35 interface for leased-line connections. You can use the Shiva Integrator 500 for backup bridging via ISDN.

TIP The Shiva ISDN Test Facility lets you test the connection to the ISDN exchange, test the connection through the exchange to other Shiva Integrator routers, and receive vital diagnostics during setup.

Mobile/Telecommuting: AccessPort

Shiva's AccessPort is a router with configuration software, which runs on Microsoft Windows 95 and Windows for Workgroups operating systems. It includes two analog ports.

AccessPort supports simultaneous IP and IPX routing and bridging and IP multihoming. It allows aggregation of two ISDN B channels using MLPPP for 128 kbps throughput and uses bandwidth-on-demand techniques.

Ordering Information

For more information about Shiva products, contact:

Phone: 800-977-4482 or 508-788-3061
Fax: 508-788-1539
World Wide Web: http://www.shiva.com/

Cisco Systems Products

Cisco provides a range of products for implementing an RLA system. The ones we cover here are the switched and fast Ethernet, PC card, router, gateway, and remote access server products.

High-Performance LAN: CiscoPro
EtherSwitch and FastHub Products

Cisco's switched and fast Ethernet products include:

- CiscoPro EtherSwitch Workgroup Switch products, which range from 5 to 25 switched ports, with high-speed expansion slots to a stackable switch that contains 16 to 192 switched Ethernet ports per stack (up to 4.8 Gbps total stack bandwidth)

- CiscoPro EtherSwitch Desktop Switch products, with 25 switched Ethernet ports and two switched 100BaseTX ports

- CiscoPro FastHub 100 products, which are stand-alone 100BaseT repeaters with 100-Mbps performance and a choice of 4, 8, 12, or 16 ports

Internet Access: CiscoPro PC Cards, Routers, and Gateways

The CiscoPro CPA200 Series internal PC cards are a solution for individual PC users. They provide up to 128-kbps throughput via MLPPP and also support dial-on-demand routing (DDR) to bring up the ISDN line only when required.

The CiscoPro CPA750 Series ISDN access routers are a multiuser ISDN access solution for small offices and home offices. They also provide up to 128-kbps throughput via MLPPP (512 kbps with compression) and support DDR. The CPA753 model has an analog telephone port.

The CiscoPro CPA1000 Series access routers are suitable for small branch offices or businesses. Like the other solutions, they provide up to 128 kbps throughput via MLPPP and DDR support.

The CiscoPro Internet Junction Gateway is a PC-based, IPX-to-IP gateway for NetWare networks. It includes a security firewall for IPX-only LANs.

The CiscoPro CPA1100 Series routing systems are LAN-server-based solutions for small- and medium-sized businesses. They are integrated with Microsoft Windows NT or NetWare servers.

Interoffice Connectivity: CiscoPro Access Routers

Along with the CPA1000 and CPA1100 Series described above, all Cisco's CPA2500 and CPA4500 series access routers support multiple WAN technologies via synchronous ports. Access lists provide internetwork security to prevent unauthorized data access.

The CiscoPro CPA2503 access router is designed for branch or main offices; the CPA2516 is designed to be used as a branch or main office integrated router/hub; and the CPA4500 is designed for central offices. They all support IP, IPX, and AppleTalk protocols, as well as frame relay, X.25, and Switched 56 WAN protocols.

NOTE All CiscoPro access routers can be managed by CiscoVision, Cisco's Windows-based network management system. CiscoVision is based on SNMP and provides a set of network management tools.

Mobile/Telecommuting: CiscoPro Remote Access Servers and Client Software

The CiscoPro CPA2509 and CPA2511 models feature integrated routers and remote access servers. They support up to 16 asynchronous modem users at up to 115.2 kbps each. Users can work from PC, Macintosh, Unix, or asynchronous terminal systems.

The CiscoRemote PC client software is a dial-up application suite for Windows-based PCs. It includes a TCP/IP protocol stack with full NetBIOS support.

Ordering Information

For more information about CiscoPro products, contact:

United States and Canada: 800-GO-CISCO (462-4726) or 408-526-7209
Europe: 32 2 778 4242
Australia: 612 9935 4107
World Wide Web: http://www.cisco.com/

Who Supplies the Equipment?

If this is a company or corporate environment, your company's network administrator will specify the equipment. If you are in business for yourself, you'll need to acquire your own equipment for telecommuting or RLA connections. The questions you are concerned with here are:

- What is the formal or informal policy for RLA or telecommuting?
- What is the standardized equipment and procurement?
- Who pays for equipment, including installation and maintenance?
- Who pays for the ISDN service—installation, wiring, usage and monthly charges?

What Your Company Provides

Your company (or client, if you have your own business) may provide the equipment. Check with your network administrator to determine what equipment you will get, and whether the company will also pay for your ISDN service. Most companies that support RLA or telecommuting require that you use standardized equipment.

What You Provide

Whatever equipment your company (or clients) does not supply is the equipment you will need to provide. You will need to check with your network administrator or client to determine what type of equipment they require or recommend.

It's possible that your company has a discount-plan agreement for your required equipment with a particular vendor or retail outlet.

Also, you may find that the equipment is provided (to ensure that it is standardized), but you must pay for the ISDN service, including all usage charges.

If you need to provide all your own equipment, check with your local carrier for any package or program that your company will support. You will see some of these packages on the retail shelves.

What Information Do I Need from My Provider?

The last sections of the RLA worksheet cover the information you should receive from your local exchange carrier, equipment vendor, and network access provider or the LAN's system administrator.

Local Exchange Carrier Information

Be sure to write down all the contact information for your local phone company. Also record the installation date scheduled and the service order number you've been assigned.

Even if your company is paying for your ISDN service charges, the administrator will probably want you to know the rates and usage charges. Record the monthly charges, per-minute rates during certain times of day, and any additional charges.

Fill in the Installation Information section of the ISDN worksheet if your phone company assigns a separate contact for ISDN installation. Even if your company is paying for the service installation, you will probably need to handle any installation problems yourself.

Provisioning Information

As we've said before, the provisioning details of your ISDN service are of the utmost importance. Having this information readily available will help you avoid many problems and to troubleshoot any problems that do arise. See Chapter 10 for a detailed rundown of the specific items. Be sure to confirm this information at actual installation time, since it's possible that some of the items will have changed from when the ISDN line was ordered.

Equipment Vendor Information

The information you need from your equipment vendor is the IOC for your RLA or telecommuting equipment. Ideally, you should get this information when you purchase (or are given) the equipment. If not, contact the equipment vendor (see Appendix B).

Information from the LAN Administrator

Your remote connection to a network involves the same configuration details as a connection to the Internet (which is just another network of sorts). After you are finished installing and configuring your remote access equipment, you need to configure your network access software.

You'll need to record dial-up, login, address, POP3, and SMTP information (if applicable). See Chapter 10 for details.

CHAPTER
TWELVE

12

A Guide to ISDN Setup for Videoconferencing

- Desktop videoconferencing planning

- Selected videoconferencing equipment

- Multipoint equipment vendors

- Multipoint service features

Is the Information Superhighway a form of transportation for the twenty-first century? Cash moves across wires rather than in armored trucks. Our news can be screened, organized, and delivered to us via a robot or agent on the World Wide Web. Now, instead of sending you to the urgent meeting across the country (they are always urgent), your company can save the airfare and reduce employee "wear and tear" by projecting your image to the meeting. So, relax, sit back, and wait for your ISDN videoconferencing setup to be ready. Soon the rigors and hassle of traveling on short notice to "urgent" meetings will be transformed into a short walk down the hall.

Get ready. Videoconferencing is going to change the way we do business, the way we learn, and the way we think.

As we've done in the previous chapters in this part, we're going to work our way through the ISDN videoconferencing worksheet shown in Figure 12.1.

> **NOTE** The worksheet addresses desktop videoconferencing requirements. If you have a room system, you may want to order more than one ISDN line.

FIGURE 12.1: ISDN videoconferencing worksheet

ISDN WORKSHEET	Page 1 of 3

Which Application Do I Want ?

    Videoconferencing

What Do I Have Now?

  Local ISDN Availability

☐ Is ISDN available from local phone company?

☐ Is a loop qualification needed?
    Completion date ___/___/___

☐ Confirm installation charge $_____

  Destination ISDN Availability

☐ Is ISDN available at destination sites?

☐ If yes, fill in the section above for all destination sites

Who Supplies the Equipment?

  Multipoint and Standards

☐ Multipoint vendor
    Contact name: _____
    Company name: _____
    Phone: _____

☐ Multipoint capabilities needed?
    ☐ Voice-activated switching
    ☐ Continuous presence
    ☐ Dial out and dial in

☐ Confirm multipoint charges
    ☐ Basic hourly charge $_____
    ☐ Usage charge $_____
    ☐ Additional charges $_____

☐ Types of multipoint services
    ☐ Connect to other vendor's equipment (H.320)?
    ☐ Share data files and images (T.120)?
    ☐ Use the document camera?

What Information Do I Need from My Provider?

  Local Exchange Carrier

☐ Contact information
    Phone company rep: _____
    Company name: _____
    Phone: _____

☐ Installation date: _____
☐ Service order number: _____

ISDN WORKSHEET *continued*

ISDN service charges

☐ Monthly charge $_____
☐ Per minute usage
    8 a.m. to 5 p.m. $_____
    from _____ to _____ $_____
☐ Additional charges $_____

☐ Installation information
Phone company rep: _____
Company name: _____
Phone: _____

Provisioning information

☐ Type of ISDN
  ☐ National
  ☐ Custom

☐ Digital switch type
  ☐ 5ESS
  ☐ DMS-100
  ☐ Siemens EWSD

☐ Local directory numbers
  ☐ B channel one   _____
  ☐ B channel two   _____

☐ Service profile identifiers (SPIDs)
  ☐ B channel one   _____
  ☐ B channel two   _____

☐ B channels
  ☐ Voice
  ☐ Data
  ☐ Both

Videoconferencing Equipment Vendor Information

☐ ISDN Ordering Code (IOC) _____

Destination Exchange Carrier

☐ Contact information
Phone company rep: _____
Company name: _____
Phone: _____

☐ Installation date: _____
☐ Service order number: _____

ISDN service charges

☐ Monthly charge $_____
☐ Per minute usage
    8 a.m. to 5 p.m. $_____
    from _____ to _____ $_____
☐ Additional charges $_____

ISDN WORKSHEET *continued*

☐ Installation information
Phone company rep: _____
Company name: _____
Phone: _____

Provisioning information
☐ Type of ISDN
☐ National
☐ Custom

☐ Digital switch type
☐ 5ESS
☐ DMS-100
☐ Siemens EWSD

☐ Local directory numbers
☐ B channel one
☐ B channel two   _____

☐ Service profile identifiers (SPIDs)
☐ B channel one
☐ B channel two   _____

☐ B channels
☐ Voice
☐ Data
☐ Both

Videoconferencing Equipment Vendor Information
☐ ISDN Ordering Code (IOC) _____

Additional contacts: _____

_____

_____

_____

_____

Notes: _____

_____

_____

_____

_____

_____

Which Application Do I Want?

Only a few years ago, the technology for videoconferencing cost $100,000 or more, filled a room, and required satellite hookups or expensive leased data lines. Now, better compression, miniaturization, faster processors, and a market expanding as fast as business travel costs climb—plus the falling price and wider availability of digital telephone lines (ISDN)—are bringing desktop videoconferencing, or DVC, within reach of large and small businesses alike.

As shown in the Figure 12.2, videoconferencing systems must be connected somehow. ISDN availability becomes a "two-fold" problem with videoconferencing applications. You may have determined pricing and availability in your area, but did you forget to consider the destination?

FIGURE 12.2:

Desktop videoconferencing with ISDN

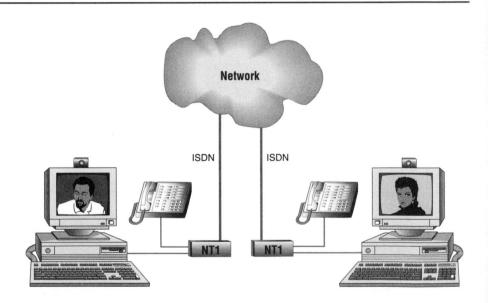

Good planning is the key. We recommend that you anticipate your needs before venturing into videoconferencing. A heads-up approach will serve two purposes:

- You can avoid potential conflicts, installation lead times, incompatibility, and other problems at your anticipated destinations.

- Multiple destinations for vendors, customers, clients, new markets, and others in your community of interest will benefit your organization.

Videoconferencing over the Internet

Until now, videoconferencing has been an application reserved for the desktops of a few power users, but several companies recently announced efforts to bring video technology to the "ordinary" computer user. These initiatives include moving live and stored video across the Internet, as well as the use of standards to create a mass market of interoperable video communications software.

Along with their proposed networkMCI WebMaker—a server with Internet connection, e-mail, and Web creation tools loaded— Intel and MCI have announced plans to work on a pilot to run video over the Internet using proposed standards. The first part of MCI's and Intel's pilot involves connecting the systems at both companies in Portland, Oregon, and Richardson, Texas, as well as Cisco Systems in San Jose, California. An outside-customer beta test is also slated.

VDOnet Corporation is running a similar pilot, offering broadcast-quality video over the Internet using the same standards. NBC and PBS will work on pumping broadcast-quality video from VDOnet's Palo Alto, California, development center for the pilot. Both of these pilots use the Real-Time Transport (RTT) instead of TCP, as well as Resource Reservation Protocol (RRP), a protocol for specifying bandwidth for network applications.

continued on next page

Although Intel and VDOnet have videoconferencing products that run on LANs, they need to test the technology's viability, because network bandwidth can't be managed as easily across the uncontrolled Internet as it can on an internal network. Intel and VDOnet are planning to ship Internet videoconferencing products next year.

Officials at PictureTel Corporation said the pilots will help videoconferencing product makers to figure out a business model for charging for the services. Although it has not announced a test product, PictureTel plans to have an Internet videoconferencing product available at about the same time as Intel.

What Do I Have Now?

As with any ISDN application, your first step is to verify that ISDN service is available in your area. For videoconferencing, you must also perform the same type of check for the destination area or areas. See Appendix A for ISDN carrier contact information.

Remember, your local phone company may need to run a loop qualification test to make sure you can have an ISDN line. Be aware that the time between when you order ISDN and when it is actually installed can range from a few days to a few months. See Chapter 14 for more information about installation delays.

Commonly Used Equipment (Authors' Choice)

Most systems provide essentially the same basic functionality. Some vendors offer low-cost, software-only solutions; others provide turnkey systems complete with computer. Regardless of its configuration, the ideal desktop videoconferencing system provides the following features:

- Application sharing
- Whiteboard
- File transfer
- Screen snapshots
- H.320 or II.324 support, and T.120 support if you want to exchange data files and images
- Two or more connectivity options
- Easy installation and use
- Multiplatform capability

Three desktop videoconferencing units that we recommend are those offered by Compression Labs, Inc. (CLI), Intel (its Proshare products), and PictureTel Corporation. Several other systems are also worth taking a look at to see if they suit your videoconferencing needs and setup. For example, Quicktime Conferencing is a similar Apple-compatible desktop videoconferencing package.

CLI Desktop Models

CLI desktop models, shown in Figure 12.3, allow you to connect point-to-point or through a multipoint bridge to any standards-based videoconference system, anywhere.

FIGURE 12.3:

The CLI desktop video-conferencing unit (top) and room conferencing unit (bottom) (*photo courtesy of Compress Labs, Inc.*)

Product Information

The CLI product offers the following features:

- Standards-based platform

- Open conferencing platform

- Industry-leading 22 fps (frames per second) video (most desk-top videoconferencing units deliver 15 fps)

- Single-board codec with ISDN interface (BRI, 2B+D, interface through inverse multiplexer board), with 56- to 384-kbps data-rate support

- Digital video camera

- Telephone handset

- CLI Desktop Video software, including file transfer, shared interactive whiteboard, on-screen dial pad, and phonebook

- Multipoint conferencing capability

- Easy-to-follow installation and user guides

- On-site installation and 24-hour telephone support

- Integrated data-collaboration software

Ordering Information

Here is the contact information for learning more about CLI's video communications solutions:

CLI
2860 Junction Avenue
San Jose, California 95134
Phone: 800-225-5254
24-hour technical support: 800-767-2254
World Wide Web: http://www.clix.com

Intel Proshare Desktop Videoconferencing

Intel Proshare Conferencing Video System 200, shown in Figure 12.4, operates on Microsoft Windows 95 and Windows 3.1 operating systems.

FIGURE 12.4:

Intel Proshare Conferencing Video System 200 (*photo courtesy of Intel Corporation*)

Product Information

The Intel Proshare unit is fully compatible with H.320 videoconferencing. In addition, the system integrates the T.120 dataconferencing standard. The integration of H.320 and T.120 conferencing standards means that you can conference with other ITU standards-based PC desktop and room conferencing systems. Or you can use standards-based multipoint control units (MCUs) for multipoint calls.

The Proshare videoconferencing system lets you conference simultaneously with customers, vendors, and colleagues in several different locations. Figure 12.5 illustrates the setup for multipoint conferencing with application sharing. You can use ISDN with a multipoint service or vendor, or with your own MCU, to share applications, transfer files, and see and hear other conference participants. Or you can simply use your LAN or WAN to conduct data-only conferences, connecting up to five colleagues across campus or branch sites. Data sharing permits simultaneous work on documents with multiple sites, whether or not other users have the application being shared.

FIGURE 12.5:

A multipoint conference with application sharing (*photo courtesy of Intel Corporation*)

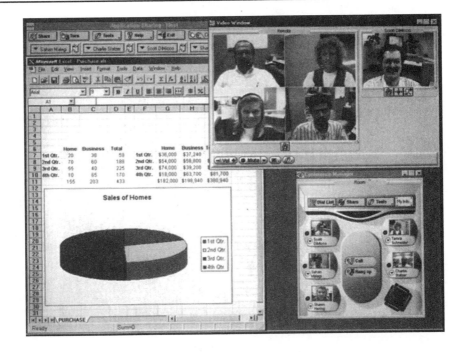

You can choose between two options:

- A switched video window that displays only the person who is talking

- A continuous presence arrangement (check with your multi-point service) that allows you to view up to four people at the same time on your PC's screen

The Proshare videoconferencing system's conference room interface and multipage notebook software feature make it easy to keep track of who's who and who's talking. This interface uses an intuitive "office view" metaphor and provides a single point of control for all product features. A Photo Exchange tool lets you store up to 75 images in a photo album, and a Business Card Exchange tool lets you trade and store electronic business cards at the end of every call.

Ordering Information

You can get information from the following sources:

- Product information by phone: 800-538-3373 (in Europe: +44-1793-431155)

- Product information by fax-back: 800-525-3019 (in Europe: +44-1793-432509)

- Authorized resellers (U.S. and Canada): http://www.intel.com/comm-net/support/anr/anrprosh.htm

- Videoconferencing worldwide availability: http://www.intel.com/comm-net/proshare/prod/atw.htm

The IOC for Intel Proshare Conferencing Video System 200 is Intel Blue.

PictureTel Live 200p

PictureTel Corporation offers several videoconferencing models, starting with a basic desktop unit and going up to a fully featured room conferencing unit. Here, we highlight the Live 200p, a single-board desktop videoconferencing system for PCs running Windows 95.

Product Information

The Live200p takes advantage of the bandwidth of the PCI bus offering high-quality, full-screen video without the need for a specialized graphics card. The system also includes LiveWare application software that integrates videoconferencing features and data collaboration functions into one interface. LiveWare includes the latest version of LiveShare Plus, which offers the following features:

- Collaborative whiteboard work
- Sharing of multiple Windows applications, transfer of files, and exchange of information through shared clipboards
- Remote control of another PC during a video call

Ordering Information

Here is the contact information for learning more about PictureTel's videoconferencing models:

PictureTel Corporation
The Tower at Northwoods
222 Rosewood Drive
Danvers, MA 01923
Phone: 800-716-6000 or 508-762-5000
Fax: 508-762-5245
World Wide Web: http://www.picturetel.com/

Other Videoconferencing Systems

Let's take a look at several other systems that may be right for your videoconferencing requirements and budget: CorelVideo, IBM Person to Person, VTEL's Personal Collaborator, and Northern Telecom's Visit Video.

CorelVideo is a videoconferencing package that piggybacks on the corporation's existing network infrastructure, without interfering or "loading" the existing network. This system offers both videoconferencing and dataconferencing capabilities. The CorelVideo package includes software and the video card. You must supply a high-quality video camera and a video overlay card. An interesting feature of CorelVideo is the application of normal phone functions, such as, call hold, call transfer, caller ID, speed-dial, and even a video phone book. CorelVideo's design, features, and cost make it a good fit for corporations considering an "internal" videoconferencing network.

IBM's Person to Person is a PC-based product that offers desktop videoconferencing using Ethernet, ISDN, or serial connections. It requires Microsoft Windows for Workgroups to utilize ISDN. The hardware requirements for Person to Person are steep; you'll need an Action Media II with Capture Option for video capture and play-back, and an ISCOM ISDN board for ISDN connectivity.

VTEL's Personal Collaborator is a single-board solution that turns a Windows 95-based PC into a full-featured desktop conferencing and communications tool. An interesting feature of the Personal Collaborator is VTEL object sharing, which lets you create an object that is linked to an application. With object sharing, you can work on a computer document with a remote colleague without slowing down the systems or compromising video or audio quality. Both users see the same object and can make changes to it, regardless of whether the application is resident on their respective PCs. Once changes have been made, the user simply clicks on the Share button to show these changes, and both users can save the object as a complete document when they are finished. The Personal Collaborator conforms to international videoconferencing standards (including H.320), so it can communicate with other standards-compliant systems from VTEL or any other manufacturer.

Northern Telecom offers Visit Video, a package that includes software, a camera, and a video board for the PC or Macintosh. When connected to another computer with the Visit Video package, the system offers real-time, black-and-white full motion video, screen sharing, and file-transfer capabilities on screen. The system needs only a single 56-kbps digital connection and conforms to the National ISDN standard. When used in conjunction with Northern Telecom's Visit Voice, the system allows users to set up a voice connection over a second channel, as well as perform other functions.

TIP　See Appendix C for resources for information about videoconferencing equipment. That appendix includes a list of resources for ISDN information, including general information, user groups, videoconferencing information, and telecommuting information.

Who Supplies the Equipment?

With videoconferencing applications, you might need to consider multipoint equipment and capabilities. You will need to deal with multipoint equipment vendors and service providers.

Multipoint Vendors

ISDN provides the capacity to conduct multipoint desktop conferences. To hold a multipoint video/audio/data conference over ISDN, participants need to be able to call into a bridge. The bridge is called a multipoint control unit, or MCU. This MCU can be inside your organization or within a service provider's network. An MCU's function is threefold:

- Recognize an industry-standard desktop conferencing application in use by the participant.

- Establish a connection with the application.

- Connect the participant with other endpoints (participants) that have called in.

An MCU is similar to the audio bridge call that you may have participated in at the office. An individual schedules a call for a particular date and time. At the appropriate time, everyone calls in and is verified to join the conference. Once a person is connected, he or she can hear all the participants.

The MCU-based conference can be any combination of audio, video, and data calls. All the participants can hear, see, and share data with each other. As each endpoint desktop conferencing system dials into the MCU, the desktop conferencing system transmits its "transmit capabilities" for data, audio, and video. After every participant (endpoint) has called in, the MCU negotiates the capabilities of all endpoint systems, establishing the conference at the lowest common denominator.

With a new multipoint system, most companies take advantage of multipoint service providers. When they become accustom to the technology, some decide to bring the capability in-house.

Multipoint service providers in the United States include AT&T WorldWorx Conferencing, networkMCI* Conferencing, and Sprint Meeting Channel. Internationally, multipoint service is available from Deutsche Telekom , TeamWorld Services, in Germany; from NTT in Japan; and from Telecom FCR in France. MCU vendors include Lucent Technologies (formerly AT&T GBCS) and Video-Server, Inc. Table 12.1 lists contact information for some multipoint service vendors, both national and international, and Table 12.2 lists contact information for some multipoint equipment (MCU) vendors.

TABLE 12.1: Multipoint Service Vendors

Service Vendor	Phone/Web Address
1-800-Video-On!	http://www.one800video-on.com/proshare.html
AT&T WorldWorx* Network Services	800-828-WORX 404-529-3099 http://www.att.com/worldworx/multip.html
Deutsche Telekom TeamWorld Services (Germany)	0-180-531-4253 (within Germany) +49-180-531-4253 (outside Germany) Fax: +49-251-910-348
networkMCI* Conferencing	800-475-5000 703-758-4550 http://www.mci.com/cgi-bin/setdoc.cgi?/ productview/content/laterals/netcon
NTT (Japan)	03-3432-3902 (within Japan) +81-3-3432-3902 (outside Japan)
PictureTel Conference Services	http://www.picturetel.com/mngdsvcs.htm
PTT Telecom NL, Multimeeting Services (Netherlands)	+31-60-22-7400 Fax: +31-20-484-3202
Sprint Meeting Channel	800-669-1235 404-649-8640 http://www.sprintbiz.com/jproducts/jconfserv.html
Stentor Conferencing Services	http://www.stentor.ca/
Telecom FCR, Multiconferencing Network (France)	+33-1-42-21-74-78 Fax: +33-1-42-21-72-06

TABLE 12.2: Multipoint Equipment Vendors

Equipment Vendor	Phone/Web Address
Lucent Technologies	800-VIDEO-GO 909-953-7455 http://www.att.com/lucent/index.html

TABLE 12.2:   Multipoint Equipment Vendors (continued)

Equipment Vendor	Phone/Web Address
VideoServer	617-229-2000 http://www.videoserver.com/htm/press20.htm
PictureTel (conferencing servers)	http://www.picturetel.com/products.htm
CLI (multipoint product family)	http://www.clix.com/Group_video/Products/multipoint.html

Multipoint Capabilities

Multipoint service providers and equipment offer three basic features:

- **Voice-activated switching:** In this type of call, you will see your-self, which is normal, in the "local" window. In the "remote" window, you will see the face of the person who is talking. As soon as the person stops talking and another begins, the remote window switches to the new speaker. When two people talk at the same time, the MCU will resolve the contention by switching the "remote" to the individual who speaks loudest.

- **Continuous presence:** In this type of call, you can see multiple conference participants simultaneously (see Figure 12.5, shown earlier, for an example). Instead of a single remote window, you will see multiple remote video windows displayed in tic-tac-toe fashion. For example, if you are one of three participants in a conference, you will see yourself in the local window, while the remote window is subdivided into two smaller windows showing the other participants in the call.

- **Dial out and dial in:** These are MCU capabilities that determine how participants in a multipoint conference will connect. Most MCUs offer "dial out" capability, which means that all the participants in a multipoint conference must wait for the MCU to call them at the appropriate time. Service providers offer dial-out capabilities because this setup makes it easier to control and maintain security. Dial-in capabilities allow the participants to dial into the MCU. Once an endpoint's access information is verified, the participant is admitted into the conference.

Multipoint Charges

When you've picked your multipoint vendor and features, be sure to verify multipoint charges, just as you would check ISP fees and phone company costs. You should confirm how much each additional multipoint feature contributes to your total charges.

Types of Multipoint Services

The following are the types of multipoint services that might be of interest to you:

- Connecting to a different vendor's equipment
- Exchanging data files and images
- Focusing on objects with the document camera

Connect to Other Vendor's Equipment (H.320)

Will you be connecting to equipment that is from another vendor than your own? If so, you will need to be sure that both videoconferencing systems support the standard video protocol H.320. This is not an issue when you connect to equipment from the same manufacturer, because you can use the proprietary scheme supported by that equipment.

Along with connecting desktop units from different vendors, the H.320 protocol allows a desktop videoconferencing unit to access a room conferencing unit. H.320 standard equipment can communicate, even when the equipment is very different in size and appearance.

Interoperability among videoconferencing units is improving. Recently, 14 videoconferencing vendors demonstrated their systems interoperating via the H.324 videoconferencing standard (a standard for analog phone service).

Share Data Files and Images (T.120)

Do you want to share files or applications during your videoconference? If you plan to exchange data, be sure to check that both your equipment and your participants' equipment support the T.120 data-conferencing standard. Without this capability, you will not be able to share applications from your local computer with other participants during your meeting, unless you all have software and hardware from the same vendor.

Use the Document Camera

The document camera (or graphic) is a popular device that is often used in a videoconferencing. It is a high-quality video camera with a lens that allows it to "see" documents. The camera hangs from a frame or gooseneck that lets the user point it at an object or document, such as a chart or map. The image is then transferred to the remote site.

TIP Using tools such as the document camera can make your videoconferences much more productive. However, as with any other type of video presentation, you should carefully plan the conference and take the time to assemble and organize the appropriate materials and equipment.

What Information Do I
Need from My Provider?

The last sections of the ISDN videoconferencing worksheet cover the information you should receive from your local exchange carriers (both local and destination), including provisioning information. The last item to fill in is the IOC for your videoconferencing equipment, which should be supplied by the equipment vendor.

Make sure to record the information your carrier's representative gives you when you order ISDN service, including the information about the person giving you information (name, company, and phone number). Of course, the installation date and service order number are also important. Then fill in the amounts for the carrier's charges: monthly, per minute, and any other charges. If you are given another contact for the actual installation of the service, record that information as well.

The Provisioning Information sections of the worksheet might be the most important ones. They determine how (and whether) your ISDN service will work with your chosen application. We've discussed ISDN types, digital switches, directory numbers, SPIDs, and B channels throughout this book, so you should have a good idea of the provisioning information your carrier should supply. Be sure to confirm this data after the ISDN service is actually installed.

PART IV

Issues and Trends

CHAPTER

THIRTEEN

13

ISDN Management
Issues

- Preparations for your company's ISDN service

- What to expect from ISDN service centers

- Network management considerations

- When to choose another network option

In the previous chapters, we covered the steps for getting ISDN service up and running for the various applications. This chapter focuses on the issues involved in managing your ISDN setup or network. This information will be useful if you are responsible for your company's telecommunications and many of your users connect with ISDN.

Many of these topics should be familiar to you by now. However, the management aspects come to play as you start to implement ISDN at multiple sites and need to address the various aspects and increased functionality of this service.

Preparing Your Company for ISDN

The success of your company's use of ISDN services—whether for online services access telecommuting, remote LAN access, or video-conferencing—depends to a great extent on the amount of support it provides ISDN users. This support may be informal or formal, but it should include guidelines for management and supervision methods, as well as funding for the equipment, service, and implementation resources.

As we've mentioned in earlier chapters, introducing ISDN adds some complexity to the equipment and network service elements for your telecommunications system. However, if you follow the guidelines in this book, your ISDN service implementation should go smoothly.

The switch to ISDN will also be easier if your company provides user education and training. By making sure that employees understand how to use the new telecommunications service, your company will benefit from ISDN more quickly. Also, standardizing the users' equipment and configurations will simplify managing these users. Equipment standardization is discussed later in the chapter.

> **NOTE** Chapter 6 includes information about developing a corporate policy for a telecommuting program. The general concepts presented there apply to using ISDN for the other applications discussed in this book as well.

Using ISDN Service Centers

ISDN service centers are available through several routes:

- Your ISDN carrier
- Terminal adapter vendors or equipment manufacturers
- Systems integrators, value-added resellers, and distributors
- Bellcore

Carrier Support Services

You can contact your carrier's ISDN service center by phone or through the carrier's Web home page (see Appendix A for contact information). These service centers can provide technical support, as well as give you details about ISDN availability. You'll be able to get technical support that relates to your PC and software configurations as well as to the ISDN service and terminal adapter. You can also learn more about using ISDN for your particular application and with your equipment. You may discover that your equipment has some useful features that you didn't know about.

The quality of the service provided by the ISDN service center varies by region and carrier. In general, you'll find carriers or RBOCs that have supported ISDN for the longest time (such as Pacific Bell, Ameritech, and Bell Atlantic) more reliable than some of the others. NYNEX, for example, has only recently opened ISDN centers, and in some cases, does not provide the support in all areas. With the merger between Bell Atlantic and NYNEX, we should see a big improvement in this area, especially considering the NYNEX goal of one million ISDN lines by the year 2000.

ISDN Terminal Adapter Vendor Services

The number of ISDN terminal adapter manufacturers who provide support centers and technical support is growing rapidly. You should take advantage of this support, especially if you must use different ISDN carriers for your sites. Vendors like Motorola, Ascend, and 3Com have technical staff to support their products and can offer the technical expertise to your staff or yourself to help implement and troubleshoot ISDN lines.

TIP

3Com has started a new program that guarantees service in 15 minutes if you use its service center for ordering and support. If you are not installed and working within 15 minutes, you can return your equipment for a refund. Contact 3Com for the specifics and requirements.

A simple test is to check some of the World Wide Web sites listed in Appendix B of this book. The vendors who have the most experience will provide many options for you to get the information and help that you need quickly. The service center should be able to simplify the ordering and installation process. By contacting a vendor's service center, you will get a good idea of what you will need to depend on if you select a certain vendor's product.

System Integrator, VAR, and Distributor Support

Systems integrators, value-added resellers (VARs), and distributors are also adding support for ISDN services and equipment. In some cases, these companies have carrier agreements to help order the service for you as you order the equipment. For example, Tech Data, Ingram Micro, and Intellicom have created support centers and marketing agreements that allow for this ISDN coordination.

Another advantage of using this type of service is that you know that any implementation and interoperability issues will have already been addressed by the distributor or reseller. If the company represents the product and the carrier, it should take care of the integration of the two for the products that it supports.

As with the terminal adapter service centers, you should check the services offered by your system integrator, VAR, or distributor. You will be able to get a sense of the experience that they have and what carriers are in their marketing agreements. If you have selected the application and the sites, but are still in the process of selecting the equipment and carriers, taking this route may help you achieve equipment and service integration. This can be an extremely effective method to set up a nationwide ISDN program with regional and local support for your remote offices.

Bellcore Support

Bellcore is another source for technical information and a general help line. As we've explained earlier in the book, Bellcore has been heavily involved in the National ISDN standard development and implementation.

Through Bellcore, you can find information about technical service issues as well as RBOC and carrier contacts. Bellcore personnel have worked with most, if not all, carriers and have tested many of the most popular applications and terminal equipment for ISDN. Bellcore will not recommend ISDN equipment, but can be helpful in other areas.

The following is some contact information for Bellcore:

National hotline number: 800-992-ISDN
Fax number: 201-829-2263
E-mail: isdn@cc.bellcore.com
Web home page: http://www.bellcore.com

> **TIP**
>
> The National ISDN Council's (NIC's) Web site at Bellcore (http://www.bellcore.com/isdn.html) features current standards information and support for end-users, manufacturers, and the industry. The NIC is a forum of telecommunications service providers, participating in Bellcore's National ISDN projects.

Managing a Network with ISDN Connections

In managing a network that uses ISDN connections, you may need to consider whether or not you need the following:

- Dial on demand
- Bandwidth on demand
- Inverse multiplexing
- Compression
- MLPPP

You also may need to deal with the setup of hub or central sites, standardization of equipment, and how to gather network management data.

Do You Need Dial on Demand?

Dial on demand is especially valuable in these cases:

- The user connects to many locations.
- The user makes remote connections only infrequently.
- The connection is the backup for a dedicated line or frame relay.

The use of ISDN as a backup for a dedicated line is becoming more common as it becomes more important to have backup facilities for mission-critical communication channels. ISDN may be the cheapest way to back up either DS0 (56 kbps) or DS1 (1.544 Mbps) dedicated lines or frame relay connections of these same bandwidths. ISDN has also become a standard interface or port on many internetworking products, which makes it easier to switch to the ISDN backup should it become necessary.

Do You Need Bandwidth on Demand?

As explained in Chapter 3, bandwidth on demand means adding B channels as the throughput or data-transmission demand increases. Intelligent software and network applications are able to take advantage of ISDN's B channels and increase the number of B channels as the application or your connection needs more bandwidth.

The use of bandwidth on demand is appropriate in LAN-to-LAN connections or internetworking environments where one B channel isn't enough for interconnection. It's also useful when an application or software program has different needs for network communications based on different types of information transfer. For example, a program may handle both e-mail exchange and patient record file transfer, which require different levels of bandwidth.

If you do implement the bandwidth-on-demand features of any application or equipment, you will need to check how these features impact the application and the user to make sure that it is effective and is providing the anticipated results. The performance of the bandwidth-on-demand applications should be monitored periodically to make sure that the full bandwidth isn't needed all of the time. If the full bandwidth is always used, you probably need to increase the amount of permanent bandwidth, and/or switch to another network service, as discussed later in this chapter.

Do You Need Inverse Multiplexing?

In an ISDN environment, inverse multiplexing allows ISDN to provide a switched network connection that appears to be larger than just one 128-kbps ISDN connection. Inverse multiplexing has been used in videoconferencing applications for years. Room video systems with inverse multiplexing capabilities can connect at full or fractional T1 bandwidth.

Ascend Communications has been one of the leaders in developing this type of use of ISDN. Many room video systems now come with Ascend Communications inverse multiplexers integrated into the units to allow multiple ISDN line connections. CLI (Compression Labs, Inc.) uses this type of arrangement to connect room video systems on a switched basis; and many others use it for the infrequent, any-to-any connections of videoconferencing users.

Inverse multiplexing for ISDN connections can get very complicated as you add sites and numbers of connections. The most common use of ISDN inverse multiplexing involves three ISDN lines for videoconferencing or T1 data network backup.

NOTE The same network architecture and technique of using multiple communications channels and inverse multiplexing to create a larger bandwidth are starting to become very popular with ATM Asynchronous Transfer Mode) networking.

Do You Need Data Compression?

There are many compression standards for data transmission, but your concern should be related to your application and the equipment that you selected. These items are what need to be compatible and interoperable in order for you to get the advantages that you want out of compression technology.

You should also realize that compression does not always increase the throughput in every case for every application or data transfer. There are cases where the compression can actually slow down transmission or increase the size of the data being transferred. This leads to two concerns:

- Make sure that all your users across the company or group have compatible compression implementations.

- Make sure that your application or software actually transmits more data with your data-compression feature turned on.

In most cases, you can turn compression on and off. This should be something that you ask prior to purchasing your equipment for your chosen application.

Do You Need MLPPP?

As explained in Chapter 3, the ability to link two ISDN B channels into one 128-kbps data stream is what the basic MLPPP (Multilink Point-to-Point Protocol) feature is all about.

Most PC users will not take advantage of this feature, because it requires a special interface or synchronous functionality that most PCs cannot support. But other network computing devices, such as routers and network servers, can use MLPPP implementations. Eventually, MLPPP will be a standard option for network-to-network connections (not end user-to-end user connections) and ISDN terminal adapter vendors will provide standard support for MLPPP. Also, most implementations will interoperate with other vendors' products, so that you do not need the same equipment made by the same vendor on both sides of the connection.

> **NOTE**   The bandwidth-on-demand feature is more useful for end users than MLPPP, because it allows for the increase in bandwidth as needed. Also, soon it will be more common for bandwidth on demand to be a dynamic or automatic feature in the software, so that the user will not need to do anything to add more B channels.

Setting Up a Hub or Central Site

When you allow users remote access to your network, they need to be able to call in and share the same network equipment. All the network users need to be connected to a central site. For analog communications, a cost-effective solution is the modem pool configuration, set up at the corporate site or at the online service provider's site. This arrangement decreases the cost and management of many individual modems at the central sites, and also eliminates the cost and complexity of supporting an individual, dedicated modem for every remote user.

When you allow many ISDN connections to a network, the same need to connect all your users to the network at a corporate site arises. You need to consider how you will aggregate these connections in one of your central sites. What does the corporate site need to have? See Chapter 3 for more information about ISDN channel aggregation.

In videoconferencing applications, the multipoint feature is popular. These applications also call for the effective and efficient central site or hub equipment configuration to connect many users to a common resource or to each other in a multipoint configuration. Chapter 3 also contains details about multipoint systems.

In the past, if you allowed ISDN connections in addition to analog modem connections, you needed to set up two configurations at your central site: one for the analog dial-in traffic and another for the ISDN dial-in traffic. This added the expense and complexity of having multiple boxes for the network interface.

Now the equipment vendors are offering equipment that integrates the analog and ISDN digital features. For example, US Robotics has expanded its analog dial-in equipment to provide for ISDN connections and support; Ascend Communications has expanded its ISDN equipment to provide for analog dial-in capability and support. Cisco Systems now offers a product to provide ISDN and analog connections at a central site. The ability to use one piece of equipment at the central or hub site greatly increases the cost-effectiveness of ISDN connections for large companies and online service providers.

> **NOTE**
> For online service and Internet connections, Ascend Communications' Max product line has been one of the most widely used for ISDN hub or access site concentration of ISDN lines. Most large ISPs use this product as the ISDN connection point in their central points of presence. Another product that is widely used by online service providers and carriers is from US Robotics.

Standardizing Your Network Equipment

We cannot emphasize enough how important it is to standardize the equipment of your users that have ISDN connections to your network. This is an extension of any PC desktop networking policy. Many companies have standardized on the office software, and many businesses agree to use standard software for exchanging files and sharing information. They found that it was too difficult and expensive to support multiple software programs and document formats. A single standard or set of standard configurations provides interoperability and reduces costs for everyone.

This is exactly what needs to be done among ISDN users and business partners using ISDN. The addition of ISDN as a network option should be an extension of the remote desktop standards for businesses and groups of users. The best type of equipment and configuration to use as your standard depends on your application. But if you do not have some method to control what equipment is used, you will run into integration and interoperability problems. This is the danger of an informal program that does not have some way to ensure that ISDN users have standard equipment and configurations.

Network Management Data

With ISDN as your basic network service, you may be wondering if there are any standard network management programs or comprehensive diagnostic network management tools you should be using. Although you can develop and track some network activity information using some of the equipment vendor's management tools, the ISDN service itself does not provide network management data or information on its own. What this means is that there are ways to create some form of network management information, but it will not be on the same level as you can get for dedicated network connections, which can supply network management data on a real-time basis.

For example, if you have an ISDN terminal adapter that supports SNMP (Simple Network Management Protocol), you can monitor some ISDN call activity. Much of the ISDN equipment for internetworking allows you to get information about the network port and connections using ISDN. You will be able to find error-recovery information and some records of port usage and uptime. The network concentration points in central or hub locations can be aggregation points for network management information and the tracking of ISDN usage. As in the analog dial environment, much of the network management functionality is in the computer network interfaces, not specifically in the network communications service.

FYI: Key Documents Published by Bellcore

Here are some ISDN specification documents available from Bellcore:

- *Features within National ISDN Enhancements Process*, SR-3681, Issue 2 (June 1996): Bellcore and the RBOCs, working with industry members and the National ISDN Users Forum (NIUF), have developed a new process for identifying feature enhancements to National ISDN. This document describes the process, presents the National ISDN enhancement candidates, and identifies their position in the process.

- *National ISDN 1995, 1996, and 1997*, SR-3875, Issue 1 (June 1996): This report summarizes the features and capabilities supported by National ISDN through the first quarter of 1997.

- *1996 Version of National ISDN Primary Rate Interface Customer Premises Equipment Generic Guidelines*, SR-3338, Issue 1 (August 1995): This document provides Bellcore's view of proposed generic guidelines for customer premises equipment (CPE) that subscribe to public network ISDN capabilities over PRI. These guidelines cover ISDN call control and supplementary services for circuit-mode and packet-mode calls, from the CPE's perspective.

- *1996 Version of National ISDN Basic Rate Interface Terminal Equipment Generic Guidelines*, SR-3339, Issue 1 (August 1995): This document provides Bellcore's view of proposed generic guidelines for terminals that offer ISDN capabilities over BRI. These guidelines cover ISDN call control and supplementary services for circuit-mode and packet-mode calls, from the terminal's perspective.

To obtain Bellcore documents, call 800-521-2673 or 908-699-5800, or order online from Bellcore at http://www.bellcore.com.

When to Pursue Other Network Options

Part of managing a network is knowing when changes need to be made. Although we've spent most of this book explaining why ISDN is usually a good choice for your network connections, there are some situations where ISDN is no longer the best option.

Low Availability and High-Usage Charges

When most of your sites are in areas that don't have ISDN available or the service is not cost-effective, this is a time to consider other network options. If only 10 or 20 percent of your sites or business partners are in areas that have ISDN, this network connection won't be cost-effective, and those users will not be able to use their ISDN lines to connect with the majority of other users. As we've said before, ISDN availability depends on your location, and is growing rapidly, especially in metropolitan areas.

A similar situation is when there are high-usage charges for ISDN (or any other dial-up service) to your sites. In that case, you may want to pursue a dedicated network option.

Fortunately, the problems with ISDN availability and usage charges will be resolved soon, as carriers strive to make ordering ISDN service go as smoothly as ordering regular analog service. Carriers are also accelerating their efforts to provide tiered and flat-rate pricing plans for ISDN service.

Need More or Less Bandwidth

Once you have the ability to connect to remote sites, you will notice the amounts of usage and bandwidth start to increase as users learn how to take advantage of this access. You should monitor the bandwidth consumed to make sure that you have not exceeded the cost or bandwidth limit of ISDN versus another network option.

Watching the usage will also help you choose the best design for your network communications. For example, if everyone calls one office for information or some data application or service, it should be obvious that this site ought to be the hub or central site of your network design and configuration. If users dial in and keep the connection up (to use an application or for other business reasons), perhaps you should change to a dedicated private line or frame relay network.

In LAN-to-LAN connections, the traffic increases between sites as new people or new applications are added. Users may find that the dial-up ISDN line is now always "nailed up" between the two sites. Higher usage cost or not, it may be time for you to install a dedicated network private line.

On the other hand, if you do not need to use all of the bandwidth provided by ISDN, or you simply have such infrequent use that multiple B channels are not required, you will find that ISDN is more than you need. When you don't make many network connections, exchange lots of data, or use multiple voice connections, you can probably get by with regular analog phone service.

Another Network Service Is Better

Your company may be using ISDN as the interim service until another service is available. For example, you may be waiting for your carrier to put in a frame relay switch or offer T1 services. When the other network service becomes available or priced more effectively for your sites and users, you will want to make the switch from ISDN.

Or perhaps your company has power users who have grown past ISDN bandwidth and are ready and trained for a new digital technology. When some of the new telecommunications technologies are finally ready, you may want to convert to one of them. For example, if and when ADSL (Asymmetrical Digital Subscriber Line) service becomes available on a more widespread basis, you might want to see if it is better compared to your existing ISDN service and charges. The cable modem is another emerging technology that you may want to consider.

You will need to analyze your needs to decide if the new services and the equipment required will be worth your time and money to make the transition from ISDN. See Chapter 2 for more information about these other telecommunications technologies.

CHAPTER

FOURTEEN

14

Some ISDN Gotcha's

- SPIDs and their formats

- Power requirements for ISDN service

- ISDN equipment's lack of interchangeability

- The correct order for ISDN setup

- ISDN installation delays

- Videoconferencing and voice call conflicts

This chapter outlines the authors' choice for the top six gotcha's of ISDN. These are not the only problems you might encounter in dealing with ISDN, but they are the ones you are most likely to come across. Every new technology had its rough spots or "bugs," and these are the ones that we selected as those most common bugs.

We have consulted many different ISDN sources and talked with the early pioneers, users, and experts to develop this list. We hope that you will be forewarned and not need to experience these problems yourself (and definitely not all of them).

We believe that the gotcha's listed here will become less troublesome, and even "non-issues" over time, but for now we feel obligated to share them and suggest how you can avoid them.

The Ever-Popular SPIDs

We've explained SPIDs in detail in this book. SPIDs, or Service Profile Identifiers, are assigned to every ISDN B channel on each line. This concept (which we have Bellcore to thank for) has caused much confusion to users and carriers alike. Although its developers imagined that SPIDs would allow for increased flexibility of ISDN lines and multiple terminals identified on one ISDN line, they have actually been counterproductive in the implementation cycle. The fact is that not many users have multiple ISDN terminals on the same line, and that is the major purpose of the SPID.

Part of the confusion about SPIDs was created by the fact that the different central office switch vendors implemented SPIDs at different times and using slightly different methods. There was a time that one SPID could be used for both B channels on one ISDN line if it were from a certain central office switch manufacturer (this is the reason many terminal adapter configuration tools ask for the central office switch type: 5ESS, DMS-100, and so on). This created different numbers of SPIDs per ISDN line. Also, some SPIDs had different formats, again based on the different central office switches.

Fortunately, the SPID problem is now getting resolved by the acceptance of National ISDN and by the work of various ISDN industry groups. For example, the VIA (Vendors ISDN Association) and the NIUF (National ISDN Users Forum) are two such groups that are aggressively pursuing efforts that will create a standard or common SPID format. This will greatly increase the ease of ISDN installation. Automatic configuration methods will require less human intervention and knowledge of ISDN.

Ascend, US Robotics, Shiva Corporation, Motorola, and Adtran are some of the companies who are coming up with some creative configuration software to ameliorate this SPID confusion for the ISDN users. They have started by putting some common SPID formats in the configuration software; and in most cases, one of these SPID formats will work. Many vendors are implementing a similar feature in their terminal adapter equipment configuration software. However, this is just an interim solution until SPID numbers are completely standardized, which should happen soon.

Know Your SPID

In the meantime, the best advice is that you request the SPID number for each line when you order ISDN service and that you document the number for future reference. Keep in mind that the basic configuration of your ISDN terminal adapter equipment will require your ISDN phone number *and* the related SPID number for each line.

The SPID is generated by the central office switch when the line is being configured for your new service request. Once the carrier has done this switch configuration, it can record the SPID and tell you the number. Make sure that you record the number in a location that can be accessed quickly. It's also a good idea to mark or identify your ISDN jack location with the SPID, central office switch type, and ISDN phone number information.

Remember, the SPID is something that the central office switch needs to synchronize with your terminal adapter equipment. If it does not receive the correct SPID, you will not be sending any data or making any calls on this ISDN line. Any service center representative, whether from the carrier or the terminal adapter vendor, will be asking you for that information whenever you call for assistance.

Formatted SPIDs

A common format for SPIDs under National ISDN is as follows:

01 *7-digit phone number* 0000

If your phone number were 555-1234, your SPID would be 0155512340000. Notice that there are no hyphens or area codes included in the SPID. Table 14.1 shows examples of SPID formats listed by service provider, ISDN type, and switch type.

TABLE 14.1:   Guess Your SPID

Provider	ISDN Type	Switch Type	Prefix	Phone No.	Suffix	SPID
Ameritech	Custom	5ESS	01	555-1234	0	0155512340
		DMS-100		321-555-1234	0, 1, 01, or 11	32155512340 32155512341 321555123401 321555123411
	National (NI1)	5ESS (version 5E8)	01	555-1234	011	015551234011
		5ESS (version 5E9)		321-555-1234	0111	32155512340111

*First and second SPID, respectively

TABLE 14.1: Guess Your SPID (continued)

Provider	ISDN Type	Switch Type	Prefix	Phone No.	Suffix	SPID
		DMS-100		321-555-1234	0111	3215551234 0111
		Siemens EWSD		321-555-1234	0111	3215551234 0111
Bell Atlantic	Custom	5ESS	01	321-555-1234	0	0155512340
		DMS-100	01	321-555-1234	0	0155512340
	NI1	5ESS	01	555-1234	000	01555 1234000
		DMS-100	01	321-555-1234	100	0132 1555 1234100
Bell Canada	NI1	DMS-100		555-1234	00	555123400
Bell South	Custom	5ESS		321-555-1234	0	3215551234 0
		DMS-100		321-555-1234	Last or last 2 digits repeated	3215551234 4 3215551234 34
	NI1	DMS-100		321-555-1234	100	3215551234100
NyNEX	NI1	5ESS		321-555-1234	0000	3215551234 0000
		DMS-100		321-555-1234	0001	3215551234 0001
Pacific Bell	Custom	5ESS	01	555-1234	0	0155512340
		DMS-100		321-555-1234	1, 2; 10, 20; or 100, 200*	3215551234 1 3215551234 2 3215551234 10 3215551234 20 3215551234 100 3215551234 200
	NI1	5ESS	01	555-1234	000	01555 1234000
		DMS-100		321-555-1234	1	3215551234 1
Southwest Bell	NI1	5ESS	01	555-1234	000	01555 1234000
		DMS-100		321-555-1234	01	321555 123401
		Siemens EWSD		321-555-1234	0100	321555 123401

*First and second SPID, respectively

TABLE 14.1: Guess Your SPID (continued)

Provider	ISDN Type	Switch Type	Prefix	Phone No.	Suffix	SPID
US West	Custom	5ESS	01	555-1234	0	55512340
	NI1	5ESS	01	555-1234	000 or 1111	015551234000 0155512341111

*First and second SPID, respectively

Power Requirements

One major difference you'll see in your conversion from analog phone service to digital ISDN service and its new interface device (terminal adapter) is the need for electrical power. The interface device for ISDN requires an electrical outlet and your power source. This means, of course, that power is needed to keep the ISDN line in service.

This is a departure from what you have come to expect of your analog phone service, especially at home. The analog service (including the ringing to signal a call coming in) is powered by the switch in your provider's central office. The power to generate the signals on your analog phone line and for the ringing of your phone is sent on the same pair of copper wires that the analog signals use for phone or fax transmissions.

This power is very stable because the phone companies have reinforced structures (central offices) with redundant power sources, usually including a gas-driven generator. This highly reliable power source ensures that you usually will have some phone service during a disaster or emergency.

Enter ISDN service with its new digital reliability and capabilities, and with a need for power. This is very much like the computer versus the old typewriter. The typewriter did not need power, and it was extremely portable—you didn't need a long power cord or nearby power source for typing. But in the end, the advantages of the computer outweighed those of the typewriter. You'll find that the benefits of the higher bandwidth, more reliable digital transmission, and other features of ISDN service outweigh the disadvantage of its power requirements.

Many terminal adapter equipment manufacturers are working on some limited battery backup features for the ISDN equipment. In fact, some ISDN equipment already has some battery or small power compacitor backup features. However, terminal adapter manufacturers are concentrating on integration and miniaturization, which conflict with adding a battery or additional power elements. Over time, this will be solved with new battery technology that will be more cost-effective and smaller.

And keep in mind, if you are using ISDN service with a computer, it's safe to say that you won't be using your ISDN service if you lose power, so the power requirement is less of a concern.

However, this power requirement has held back many local regulators and state commissions from defining ISDN as a basic service. The ability to call during an emergency has become a basic phone feature that most commissions will require of any universal service.

NOTE If you have purchased a new VCR recently, you may have noticed that the new models now have some power capacity so that you don't need to reprogram or reset the clock whenever you lose power. This is the kind of subtle development that you will see in the power area of ISDN terminal adapters. This power requirement will be part of every new high-speed digital service, including the emerging services described in Chapter 15. This means that the need for backup power will continue to increase, and so will the speed with which the various equipment vendors provide solutions.

Non-interchangeable ISDN Equipment

Many ISDN service centers get calls from users asking why their ISDN line doesn't work anymore after they switched to another ISDN terminal adapter brand or model.

This is another difference you will find in the digital ISDN service: you can't just swap one part for another. Since the SPID numbers need to be synchronized, and various terminal adapters generally require a different configuration on the ISDN line, exchanging ISDN terminal adapter equipment on the same ISDN line usually won't work.

The good news is that ISDN terminal adapters are getting better and can support more and more applications. This minimizes the need for swapping equipment.

However, if you do need to move ISDN equipment from one line to another, or replace one terminal adapter with one from another vendor, you will probably need to reconfigure your equipment and also have the ISDN service provider perform some configuration on the ISDN line itself. On the other hand, you might get lucky and find that your existing configuration will work with another terminal adapter.

Interchangeable Analog Communications Equipment

In the analog phone world, equipment manufacturers have realized that there is only one analog phone interface, with one technical set of characteristics. But analog phone equipment wasn't always interchangeable.

In the early days, after the breakup of AT&T in 1984, some customers purchased "cheap" phones that would not ring or could not send the right dial-pad signals to make or complete calls on the phone networks. Some phones actually partially melted from the ringing power sent. It took some time, but now we expect any analog phone device to work on any analog phone service.

When modems were introduced, one of the reasons that they were not readily accepted is that they had some nonstandard software or hardware features that created "random" connections. Now modem equipment is standardized, although there is still some modem software that can be unreliable.

Digital interfaces are not so kind or forgiving. And the flexibility and variety of ISDN configurations can be confusing. But the manufacturers are beginning to design ISDN equipment that is interchangeable with other makes and models. The standardization that allows this includes ISDN ordering codes (IOCs), standard SPID formats, and intelligent software in the terminal adapters.

When ISDN terminal adapters are truly interchangeable, you will see things like telecommuting centers with guest ISDN lines and hotels with business suites that include ISDN service.

Does the Equipment Match the Application?

Buying your terminal adapter or ordering ISDN service and then deciding what to do with the service is like buying a PC and then deciding what software it is going to run. If you buy your PC this way, you will soon find that the memory is too small or your PC does not support one key feature of your software application. After you buy the equipment and install the ISDN line is not a good time to decide that you don't need the ISDN service or that the equipment is too expensive or complex.

As we've said before, our best advice for people ordering ISDN service is to work backward:

1. Pick your application.

2. Select your equipment.

3. Make the service connection.

Of course, you still need to verify that ISDN is available in your area before you do anything else. But once you know that ISDN service is available, you should follow the order above. You need to start with the key application first, then find the terminal adapter that supports that application, and then order the line. The application and terminal adapter dictate the ISDN line configuration.

Another big advantage to following this order is that, by selecting your application and equipment first, you will have more resources for support as you go through this process:

• The support for your application

• Equipment resources, which may include ordering support

• The carrier or RBOC, who offers additional support for the service itself

For example, you can get a wealth of information from terminal equipment vendors' technical support centers and customer service centers. The service centers of carriers and terminal adapter manufacturers will ask you to clearly identify your application, your PC or terminal configuration, and your ISDN terminal equipment.

This order also provides an easy trail to follow to identify the total cost of using or converting to ISDN. By discovering what you need for your application, you can analyze the total cost prior to buying any equipment or ordering ISDN service.

Availability and Installation Interval

As we have stated in many places in this book, the overall availability of ISDN has greatly improved over the last two years. It is being quoted as 80 to 90 percent availability in major metropolitan areas across the country, and some carriers or RBOCs talk about 95 percent coverage in some areas.

Well, the reality is that those numbers are averages, and you may not be in an area where ISDN is available, or it may be available only by "extending" it to your home. It may be extended from another central office or extended because of the distance from your home or office to the phone company's central office.

The result is variability in ISDN availability and in how long you will wait for installation. If ISDN is readily available in your area, you may need to wait only two days for the ISDN line to be installed. The average length of installation is usually a week to ten days.

But in other cases, the interval between ordering and installation is very long. Even Pacific Bell has had cases of switch overload or blocking (where the ISDN capacity has been met), and the new switch equipment required was not available for months.

If your location is far from the phone company's central office, this can cause installation delays, as well as some additional costs. Some carriers charge for the extended mileage equipment and labor. If you live out in the hills, you might wait three to four weeks for ISDN installation due to the extra engineering and ISDN BRITE (Basic Rate ISDN Terminal Extension) cards required for the longer distances.

You might find out that you need to wait a couple of weeks or months, even after you have checked for availability on a Web home page or through an 800-number service center. Your source might not have had up-to-date information about the central office switch capacity or your exact location's distance from the central office. This variable timing can play havoc on a large regional or national project. Scheduling can become somewhat of a nightmare, and you can't plan on availability all at one time.

Placing an order is one sure way to validate service and determine when the installation can be done. The safest thing to do is to order the ISDN line with some lead time and try to get some commitment from the carrier as to the installation date.

Many ISDN service providers are aware of this problem and are working to increase the ability to quickly answer the installation question and reduce the number of situations that create delay. This will become less and less of a problem as all carriers increase ISDN coverage and refine the new equipment that improves the extended ISDN technology.

Videoconferencing and Line Conflicts

Using ISDN as your primary voice and data connection may have some inherent limitations. You cannot have a videoconference going while you are trying to send and receive voice calls on the same line used for your customers. This can be a problem when your customers need immediate access to the voice line and you want to use the same line for videoconferencing. It would be difficult to put your customers on hold while you finish your videoconference.

Although you can add more lines, a situation like this may indicate that the ISDN service is not being used correctly. ISDN cannot support multiple lines for a small business when those same lines will be used for videoconferencing.

This is no different than using one line for phone calls and fax calls at the same time. You will run into conflicts in trying to use the same line. On the other hand, you can use one B channel for voice transactions or calls and one B channel for data applications or online service access. This is the type of simultaneous use of ISDN service that is cost-effective and efficient.

A Note about the Gotcha's

These are the most common gotcha's for ISDN implementation, and they cover most causes of installation problems for new users and systems. But you'll soon forget about any of these troubles when your service is up and running and you can enjoy the benefits of ISDN.

We hope that this will be a much better experience than your first modem configuration session or setting up your first PC. And remember, all the problems mentioned here will be less troublesome or eliminated completely as the various industry players work through these items during the next year.

CHAPTER
FIFTEEN

15

Future Trends
for ISDN

- Some trends for ISDN service in the United States

- Some trends for ISDN service in other countries

- New applications for ISDN connections

- Emerging telecommunications technologies

As you already know by now, the future of telecommunications is in digital, high-bandwidth connections, and this is the type of service ISDN can provide. During the next three years, interest in ISDN will increase. More equipment vendors will offer ISDN equipment, more providers will offer ISDN service and equipment packages, and getting ISDN service will be easier. ISDN service installation will also become simpler, with the combination of the service standardization (including the SPID), easier configuration of terminal adapter equipment, and more support from ISDN carriers.

The four applications we have highlighted in this book will continue to grow and new applications will emerge. They will be used with many network options, but ISDN will continue to be one of the most popular.

This chapter discusses some of the future trends for ISDN, many of which are happening as you read this book.

The Future for ISDN

As we see a rapid increase in the worldwide availability of ISDN, there are certain trends that will continue and some new trends that will develop:

- An ISDN interface in most PC units with analog modem functionality

- ISDN service access as the recommended option for all Internet and online service providers

- Complete standardization of ISDN service

- New technologies benefiting from the "ISDN experience"
- ISDN packages available from retail outlets
- A drop in the price of ISDN equipment and service

ISDN as the Digital Modem in Your PC

Here's our prediction: Within the next year, the ISDN interface will be reduced to the term "digital modem," and it will be included in all the equipment that today has a built-in analog modem—this means not just standard PCs, but any terminal or network-connected device that uses a modem today. For example, many credit card terminals, routers, and fax modems will have ISDN as a standard option in the next year.

As the ISDN terminal adapter technology matures, adapter components will become smaller and cheaper to produce. Many PC manufacturers will include ISDN equipment in their "box." The built-in ISDN terminal adapter will come with simple software tools to configure the PC for ISDN. But you probably won't need to use them, because the configuration will already have been set up for you, just as with your built-in analog modem. Microsoft is already working on giving Windows 95 the same self-configuration capability that is currently provided for analog modems.

A single card in your PC will handle both analog and digital connections for you. The software will manage whether it is a digital or analog remote connection, just like the current communications software decides which speed your modem will use for a particular analog connection.

Of course, the ISDN service carriers will need to do their bit to make ISDN attractive to PC manufacturers in order for ISDN to become your digital modem. They will need to make sure that ISDN service is as easy to use and install as analog service. If this does not happen, PC manufacturers and others who support the new ISDN digital interface will not support ISDN for long, especially if it increases their cost of distribution and service. The PC manufacturers are looking for the right technology for the "networked" computer of the future.

Remote Computing and the Networked PC

The explosion in remote computing is a major force driving the growth of ISDN. Along with your ISDN "digital modem," soon you will see a set of remote computing applications packaged in every PC, with many other networking options. Remote workers require quick-and-easy access to information and a way to share information with others.

In future client/server architectures, your application and database will be some combination of local and remote resources, hidden from the end user. Applications will be divided into smaller applets that are shared within a network, and users will not need complete applications as they do today.

So what is the "networked" PC of the future? One view is that it will become a smaller, low-cost terminal that can connect to the major resources of a service provider, institution, or company.

Another view is that the networked PC will become a more powerful workstation that requires minimal connectivity to other corporate or company resources. However, we believe that this is less likely to happen, because databases and other resources are dynamic and require frequent updating; keeping real-time information available to multiple users is beyond the capabilities of one workstation or PC user.

But either way the networked PC evolves, ISDN will be a popular method for network connectivity.

ISDN, Internet, and Online Services

ISPs were one of the first major online service providers to package ISDN with their service. They were also one of the first to subsidize the ISDN terminal adapter hardware to reduce the cost of converting from analog modem to digital ISDN service for access to their services. Now some ISPs are the first to start pricing the digital ISDN access the same as analog modem access.

Internet and online service access alone may prompt the deployment of millions of ISDN lines each year for the next three years. And ISDN does not need to be the most popular or common network access option to reach that number of lines.

There are forecasts that ISDN deployment will exceed over 10 million lines nationwide by the year 2000. The online service applications can contribute from one-third to one-half of that volume of lines if the current indications and trends continue.

After the year 2000, there will be other network options, but ISDN will still be around, possibly as the default option for high-volume consumer access. This may relegate ISDN to the lower-tier, "power-lite" user option for online services.

NOTE Keep in mind that the Internet currently cannot support individual user access at faster than 200 kbps. This limitation stems from the Internet infrastructure, server communications, and the TCP/IP protocol, which are all integral parts of Internet service today.

ISDN Service Standardization

You might think that we already have a standard service called ISDN, but today the services still have some differences. Our prediction is that these will be distilled to just one or two options for ISDN service within the year. As with regular analog phone service, you will not need to know the switch type or software version of that switch to order or install your ISDN service.

All the industries involved—Internet and online service providers, local phone companies, national carriers, terminal adapter manufacturers, and switch vendors—need to agree on the features and customer applications that they will support. In fact, carriers will be plagued with high service costs and service center chaos if they don't have the discipline to standardize ISDN. They need to identify the volume applications and create "templates" of standard provisions and user support for these applications.

> **NOTE**
>
> Bell South, with Bellcore's assistance, has introduced the EZ ISDN program, with templates for standard ISDN equipment configurations. Motorola and Bay Networks already offer equipment that is compatible with the EZ ISDN templates.

The goal is a completely standardized network interface and standard configurations for specific ISDN applications. A bonus will be the development of interchangeable ISDN terminal adapter equipment that conforms to these standards.

ISDN Paves the Way

As other digital, higher-bandwidth network options are developed and made available, you will see that the ISDN service introduction has provided invaluable experience to all those concerned. The industry has learned much about developing compatible terminal equipment, defining a telecommunications service, standardizing the service, and developing communication channels.

As ISDN gets more attention and market share, other communications access methods will also thrive. ISDN is not the only option for these applications and never will be the only choice.

ISDN will become somewhat of a de facto standard interface for some applications, similar to Ethernet in today's LAN environments; many PCs will come with a network interface card for Ethernet connections, whether it is requested by the prospective purchaser or not.

ISDN will be one of the WAN standard interfaces, in addition to analog dial. In the future, Ethernet and frame relay interfaces will become more available and also be built into the PC and network hardware devices.

ATM will have increased popularity and growing deployment. Its availability will also increase with more carriers and ISPs using ATM as the backbone of frame relay networks and Internet services. Frame relay and ATM both represent the extension of the Broadband ISDN (B-ISDN) standards. This is why many see ISDN as the basic digital network that can be migrated to these other networking options.

Internetworking and hardware vendors will move on some of these new technologies faster than they did with ISDN. For example, terminal vendors and equipment manufacturers are now working together to develop and test ADSL, cable modems, and wireless cable services. Also, RBOCs and other carriers will move more quickly to standardize the new emerging services and invest in the terminal equipment required.

In the end, these services and the required support infrastructure will have the benefit of the "bumpy" road of ISDN introduction over the last ten years. The developers of these new services will have the benefit of this early experience, and this will help them to get the support and infrastructure in place. Carriers, equipment vendors, and service providers have also learned how important it is to have technically literate support personnel available when they introduce new digital network services.

ISDN Goes Retail

To go along with the other trends, you will soon see more ISDN packages being sold at retail channels. As ordering and installing ISDN service becomes more simplified—which is happening now thanks to IOCs (ISDN Ordering Codes) and National ISDN—and ISDN interfaces are integrated into PCs and networking devices, the retail outlets will jump in with ISDN packages. Customers will be able to access ISDN service centers, run by carriers and equipment providers, 24 hours a day, 7 days a week, so they can get the support and troubleshooting advice they need.

Electronic ordering will not be far behind the retail channel acceptance of ISDN. Some of this work is being done with the early success of Microsoft's Get ISDN program.

> **TIP**
>
> Microsoft's Get ISDN Web site is at http://www.microsoft.com/windows/getisdn/. See Chapter 10 for more information about the Get ISDN program.

The retail outlet managers will be very cautious, however. If the carriers and equipment manufacturers do not reduce the complexity of ISDN service setup and address user confusion for retail environments quickly enough, the retail market may be dissuaded from presenting or reselling ISDN equipment. If the product doesn't sell, it won't matter whether the equipment is integrated or not. A high volume of returns kills retail products.

ISDN Equipment Costs Decline

The price of ISDN chipset and terminal adapters is dropping as the number of features are increasing. ISDN terminal adapter equipment manufacturers are enhancing their product lines and providing increased support for ISDN installation.

More vendors are entering the market on a national and international basis. For the next couple years, we will see ISDN equipment prices continue to fall, and improved functionality. After that time, there will be some consolidation and fallout of the vendors who were not able to acquire sufficient market share to survive.

At the end of last year, the price for some ISDN terminal adapters was dipping below the $300 per unit mark. You could even find some equipment that offered limited functionality for basic ISDN applications, and equipment that came with a few subsidized Internet and RBOC programs, for below $200; a very few programs offered ISDN equipment for below $100.

Soon you will see full-featured ISDN terminal adapter equipment priced below $200. With various carrier, RBOC, and Internet packages, many programs will offer ISDN equipment at or below $100.

The equipment vendors have been working on the integration of the ISDN terminal adapter function into various computer devices (including the PC and network servers). As this becomes more common, within the next two years or so, the stand-alone ISDN terminal adapter will be phased out.

Carrier Trends

The RBOCs, long-distance carriers, and local carriers are working to complete the productization cycle for ISDN. They will continue to simplify the pricing. You will be able to get packages with flat-rate or tiered pricing plans, so you can take advantage of the best pricing package for your usage pattern. Carriers are using the four applications we've covered in this book as the target applications for their new pricing packages and tariffs.

This will lead to a small number of packages with tiered and flat-rate pricing for the high-usage, power users. It will also provide some lower-tier packages for infrequent users who want better performance than 28.8-kbps analog modem service. If this is not sufficient in functionality or bandwidth, some users may migrate to higher-bandwidth options, but many will hang onto ISDN service for as long as possible. ISDN has a good two- to three-year head start on most of the other digital communication technologies.

Industry Trends

Many carriers, RBOCs, and equipment manufacturers are entering strategic agreements to escalate the development and availability of new digital services and applications. One major industry trend that will continue for the next few years is that of companies buying other companies for their ISDN technology. For example, Cabletron recently bought Network Express, who had just purchased IBM's Waverunner ISDN product line.

> **NOTE** IBM managers have stated that they will use the money from the Waverunner sale to focus on ATM and other network technologies. IBM has never had a large stake in network communications hardware, and that is why the Waverunner product seemed to have uneven support for product development, marketing, and promotions. We hope that the ATM product line gets more resources, support, and management attention, or we predict it will be sold in a few years also.

There will be an increase in buying rather then developing ISDN technology internally for the near future. Companies feel that the growth of ISDN is here now, and their timing in marketing ISDN-compatible equipment is critical. The three major internetworking or routing vendors (Cisco Systems, Bay Network, and 3Com) have bought what they need for ISDN and now are integrating and promoting it in their respective product lines.

The modem vendors have already decided that ISDN is the next step after 28.8-kbps modems. You will see 28.8-kbps modems by Motorola and US Robotics on the retail shelf, with the corresponding ISDN equipment next to them.

International Trends

The international deployment of ISDN continues to grow. A number of countries deploy ISDN digital service as a basic infrastructure, supported by government-owned or -subsidized public phone companies. This will increase the demand worldwide, since most industrialized nations will have ISDN service as a standard, interoperable service. ISDN may not be the most prevalent service for all countries and for all users, but it will be a common business service in most countries.

We covered international trends for ISDN in Chapter 4. Here is a brief summary of the future for ISDN in other countries:

- **Asia:** Countries in this part of the world that are making major investments in new telecommunications infrastructures, such as Korea and China, will have the most rapid growth in ISDN deployment. Malaysia and Thailand will also experience growth in all telecommunications-related areas, including ISDN services. They will be followed by other countries in that region that are starting to increase the overall availability of telecommunications in their respective countries.

- **Europe:** These countries will continue to adhere to the Euro-ISDN standard and the work that the IMIMG is sponsoring with the close involvement of the ETSI. This will lead to increased acceptance of ISDN as a European network standard. The individual standards and varieties of ISDN service throughout the countries will give way to the Euro-ISDN standard and increased interoperability between the European countries and other countries.

- **Africa:** South Africa is just starting to invest in ISDN and will take the lead in ISDN deployment in Africa. Some countries in Africa are getting behind Euro-ISDN and the European efforts to increase awareness and availability of ISDN.

- **Central/Latin America:** The majority of the growth in telecommunications and ISDN will occur in Argentina, Brazil, Chile, and Mexico. These countries will invest in other transport technologies as well, but ISDN service will be a basic component of any new technology that is deployed.

- **Australia:** This country has been very busy improving the network infrastructure and support systems, and buying some of the best network components, systems support, and management expertise. Australia is working to provide a state-of-the-art network platform for the major business and residence centers of the area. There will be many network upgrades and new projects in Australia, and ISDN will be one of the common services offered (along with other data network services, such as frame relay and ATM). Along with China, this will be one of the largest growth areas in the telecommunications market overall.

Future ISDN Applications

With ISDN's widespread availability and ease of use will come more and more applications that can benefit from ISDN service. The following sections describe some of these up-and-coming applications.

Telemedicine

Currently, ISDN service is being used for some telemedicine consultative applications. For example, it might be used when a second opinion is required and the use of ISDN as the video connection is an economical alternative to having doctors travel or personally attend to a patient that has already been diagnosed. ISDN service is not appropriate for higher-bandwidth telemedicine applications that require live action and broadcast-quality video definition for medical operations, such as remote surgery.

When ISDN service is available in more rural areas, this will become a very cost-effective way to improve health care and decrease costs in those regions. This is especially true as the PC manufacturers start to include some of the video and ISDN network capabilities "in the box."

NOTE Some health professionals have been trying to get simple health-care eligibility and verification to go electronic, but it will take an overhaul of that industry's financial model for this to happen. Even simple electronic claims processing requires a standard format. Of course, this aspect of health care would benefit from a faster network communication method, such as ISDN.

Customer Service

Some companies are already considering using ISDN connections to offer better support for their customers. For example, a software company might want to provide technical support with the video and data-collaboration capabilities of an ISDN-connected desktop videoconferencing package. Rather than sending programmers and other technical staff to customer sites, which is expensive and makes the experts unavailable to other customers, the company could bring the support personnel and clients together in a real-time videoconference. The customers are satisfied, because the ISDN connection reduces the company's response time and gets them to the right expert more quickly.

Of course, companies will implement ISDN services for their top-tier customers at first, until it becomes cost-effective for their other customers. This concept can also be extended to include a means to share product news with customers more personally, using ISDN connections to send messages.

Home/Remote Banking Services

Mellon Bank, with Bell Atlantic and PictureTel, is starting to deploy video banking stations using ISDN and PictureTel video equipment. There are other bank trials using video for loan applications or quick turnaround of loan officer approvals for customers from remote branch locations. These services could be extended to allow customer home connections for banking services, including loan applications. In fact, Mellon Bank intends to extend the current stations to include customer access from home.

Remote banking services expand a bank's accessibility beyond traditional hours and regions. This application will become more common as more banks merge and electronic communications are improved.

Distance Learning

Distance-learning applications allow for training, education, and instruction from many remote sites. Schools have begun to implement this type of technology for communications among remote experts, libraries, and school interexchange programs.

As mentioned earlier in this book, distance learning also applies to the business environment. Pacific Bell uses ISDN service for some PC-based instruction for employees. This application not only provides the multimedia information format, but also updates the records for the employee to record and document this coverage. The

ISDN service allows for quick access to information and real-time updates to the human resource database files.

Many other businesses are using distance-learning, broadcast technology to share new product information or company announcements for remote sites that are not connected to the corporate LAN. Using distance-learning applications for training can dramatically reduce travel time and costs.

Point of Sale

How many times have you waited for a clerk to dial into the bank for credit-card clearance? This is happening more as the dial traffic increases and the modem technology cannot handle the larger amounts of information required for credit-card transactions. These "point of sale" applications could become very popular with retailers during the Christmas sales rush. Restaurant owners would also like the transactions to happen faster for quicker customer turnaround.

Verifone provides credit-card-transaction terminals for many retail merchants in the world. The small, gray credit-card terminal is one of the most popular units in the country. It has software to read your credit card, send in the number for verification and authorization, and then complete the transaction with the signature printout and final settlement. Multiple banks and financial institutions may be involved. This has become the standard electronic banking method for many retailers.

Verifone tested ISDN as a network interface for the credit-card terminal in 1995. Recently, it announced a national program called Digital Xpress for retailers who want to migrate to the faster, more reliable, digital service for their electronic banking.

Verifone also plans to integrate the ISDN terminal adapter with analog conversion jacks in its new terminal models. This will allow the retailers to reuse their analog equipment (such as phones and fax and answering machines) and increase the throughput of their transactions. This program is designed to handle a high number of transactions per day, but will allow for the retailer's normal and holiday volume to be handled more efficiently than with analog dial service.

Home Office Personal Phone Navigator

Some vendors are starting to market small, "magic" boxes based on ISDN service. These boxes include a package of voice services along with data services for your PC connectivity needs.

A system like this could become your "mini-PBX" or phone system for your home. You could provide different numbers for your home and your work. It would have different messages for each number dialed. Your faxes could be stored per your instructions dialing into the same equipment.

A company called Jetstream Communications, located in California, is conducting trials of this type of equipment as we write this book, and it may be available by the time you are reading this. Other companies will soon follow with similar products.

Government Services

Driver's license applications, car registrations, voter registrations, lottery network access, employee training, and public announcements are all public service applications that can take advantage of ISDN.

Many government agencies are already using ISDN internally for their own communication needs. Expanding its use to public services would make it easier to support customer requests, applications, and registration activities that require simple form completion and submittal. This could be combined with an online service or Internet connection to provide 24-hour access to these basic services.

Another Look at Emerging Technologies

As we said earlier, the introduction of ISDN has paved the way for other emerging network technologies. Some will be viable network access alternative services for the applications that we have highlighted in this book.

As you would suspect, the leading application driving many of the network communications today is the Internet. Internet access and the backbone network for the Internet are the reasons users and carriers are demanding higher and higher bandwidth.

Many of these new services will mature and become products much faster than ISDN has. The competition within the telecommunications industry requires that the pace of introducing services and related support is twice or three times as fast as in the past. New services that used to mature in two-year project cycles must now be ready in one year or less.

In some cases, these new services will be product extensions of existing application packages and marketing programs that have already been developed for ISDN service.

These services could be categorized as slow (analog dial), fast (digital ISDN), faster (cable modems and ADSL), and fastest (ATM), with many other services fitting into the bandwidth spectrum. Realize that ISDN service was the *first* digital subscriber line (DSL).

We covered the other telecommunications technologies in Chapter 2. Here's a brief summary of the future for these technologies:

- **Cable modems:** People are interested in cable modem service because it sounds like a flat-rated service at some "shared" Ethernet speed to the home. The key word here is *shared*—the real throughput cannot be determined until the existing cable plant is upgraded and the architecture is defined, so that the providers can tell how many users are sharing the Ethernet bandwidth. This will define the actual usable bandwidth and security of this service. Many cable companies and operators are trying to create a standard offering for the residential market. They are also now starting to think about the business and work-at-home market. The big question here is: Do you want the cable company providing a data service with your cable service, possibly with telephone service combined (if they can actually do it)?

> **NOTE**
>
> TCI with @Home is trying to create a larger-scale offering of cable modem service, with the cooperation (and equity ownership) of Comcast and Cox. This service has been delayed multiple times. Motorola's Cablecomm Data Products, Zenith, and LAN City have cable modem equipment that has been or is part of this early beta and trial environment.

- **XDSL (the ADSL/HDSL/SDSL/RADSL/VDSL family):** This is the next wave of DSL. All these technologies offer three channels on the same copper pair(s) of wire connection. Generally, these are a high-speed channel (downstream), a slower duplex (upstream) channel, and a third channel for the normal analog phone service. The bandwidth for the first two channels varies by the type of XDSL and the specific vendor implementation. Eventually, the service will provide one to two megabits upstream and up to six megabits downstream for a robust, full-duplex, reliable, and dedicated access digital service. There are new versions of this technology expected by year-end.

- **DirectPC:** This is a new PC satellite service for network connectivity service that is a spin-off of the DirectTV cable TV product, which has been very successful recently. Current availability seems to be very low, but it has some great possibilities if it can be made available prior to any other land-line-based digital service.

- **Wireless Data:** There will be an explosion of the wireless data networks and options. With cellular and PC networks growing and on a collision course for market share, so too will the wireless data market go. In addition to those larger networks with some large regional and national reach, there will be other wireless data options, including RF (radio frequency) technologies and a large variety of wireless hybrid networks. However, users are looking for Ethernet-like (10 megabits) connectivity, and many wireless data services are hard-pressed to provide a reliable 19.2-kbps connection. If this speed does not improve, wireless data will be relegated to a niche data service offering for large corporate users and some specific applications that demand short, "bursty" transactions.

Where Does ISDN Go from Here?

We envision a PC with the intelligent software to take full advantage of all three channels of Basic Rate ISDN. While you sleep, your machine will be able to pick up your e-mail, favorite Web page updates, voice mail, faxes, and news line information. This is because it will be programmed to take advantage of the best time for accessing resources (off hours) and for the least usage charges from your phone company or service bureau. You may even be able to have a personal organizer that you can plug into the ISDN line at night to do the same type of activities.

With the reduced cost of the ISDN equipment, software applications (possibly freeware), and service, you could easily use ISDN connections for family video calls to reduce your travel and keep in touch with broadcast-quality pictures. The hardware, software, and respective video algorithms could support this by then.

ISDN will continue to mature as a product or service within the next three to four years. There will be other opportunities for product extensions with new applications or repackaging the service, but the next three to four years should be the peak of its product life cycle. ISDN service will eventually be replaced by another wave of digital technology that can be deployed as ubiquitously and economically. Could this be cable modems, frame relay, or ATM?

APPENDIX

A

ISDN CARRIERS

This appendix lists contact information for ISDN carriers, beginning with hotline phone numbers for the seven RBOCs, followed by national carriers and international carriers.

Regional Bell Company Toll-Free Numbers

RBOC	Phone
Ameritech	800-832-6328
Bell Atlantic	800-570-4736
Bell South	800-428-4736
NYNEX	800-438-4736
Pacific Bell	800-472-4736
Southwestern Bell	800-792-4736
US West	800-246-4736

National Carriers

The following national carriers are listed here:

- Ameritech

- AT&T
- Bell Atlantic
- Bell South
- GTE
- Northern Arkansas Telephone Company (NATCO)
- NYNEX
- Pacific Bell
- Southwestern Bell
- US West

Ameritech

Contact	Address/Phone
Home Page	http://www.ameritech.com/
Data Services	http://www.ameritech.com/products/data/
Team Data Support Center	http://www.ameritech.com/products/data/teamdata/
E-mail	ameritech.teamdata@x400gw.ameritech.com
Phone	800-832-6328; outside U.S.: 847.248.8093
ISDN Direct Service	http://www.ameritech.com/products/data/rates.html

Contact	Address/Phone
Phone	800-419-5400 (Home Professional & ISDN Service) 800-417-9888 (Business Professional & ISDN Service)
A Guide to Business Solutions	http://www.ameritech.com/products/business/
Ameritech ISDN	http://www.ameritech.com/products/data/isdn/index.html
Ameritech ISDN Direct Availability	http://teamdata.aads.net/isdn/quote.htf

AT&T Digital Long-Distance Service

Web: http://www.att.com/home64/

Phone: 800-820-6464 (Residential)
800-222-7956 (Business)

> **NOTE**
>
> Before you can order AT&T Digital Long-Distance Service, you must first place an order for an ISDN line from your local phone company. Most local phone companies have established 800 numbers and information databases to assist you in determining ISDN availability and in ordering ISDN service.

Bell Atlantic

Contact	Address/Phone
Home Page	http://www.bell-atl.com/
ISDN-Speed	http://www.bell-atl.com/html/hot/
Small Business ISDN	http://www.bell-atl.com/isdn/sbs/
Small Business ISDN–Pricing and Tariffs	http://www.bell-atl.com/isdn/sbs/text/html/order/price/price.htm
Residential ISDN	http://www.bell-atl.com/isdn/consumer/
Residential ISDN-Pricing	http://www.bell-atl.com/isdn/consumer/graphic/html/order/price/
InfoSpeed Sales and Service Center	800-204-7332
ISDN Sales and Technology Center	800-570-4736

Bell South

Contact	Address/Phone
ISDN Single Line Service (Business Environment)	http://www.bell.bellsouth.com/products-services/isdnintr-work.html
ISDN Ordering and Billing Information	800-858-9413

Contact	Address/Phone
ISDN Technical Support	800-858-9413
ISDN Availability	http://www.bell.bellsouth.com/ products-services/isdn-prod-avail.html
ISDN Individual Line Business Service Pricing	http://www.bell.bellsouth.com/ products-services/isdn-pricing-single-line.html
Single Line ISDN for Consumers	http://www.bell.bellsouth.com/ products-services/isdnintr-home.html
ISDN Individual Line Residential Service Pricing	http://www.bell.bellsouth.com/ products-services/isdn-pricing-residential.html
Data Customer Support Center	http://www.bell.bellsouth.com/ products-services/dcsc.html

And here's a "snail mail" address:

ISDN Single Line Service Support
Data Customer Support Center
1950 West Exchange Place, Suite 500
Tucker, GA 30084

GTE, Inc.

Contact	Address/Phone
Home Page:	http://www.gte.com/

Contact	Address/Phone
Business	http://www.gte.com/Cando/Business/business.html
Home Office: Turnkey Solutions for Today's Telecommuters and Entrepreneurs	http://www.gte.com/Cando/Homeoffi/homeoffi.html
Intelligent Network Services, Inc.	http://www.gte.com/Cando/Internet/Docs/Access/ins.html
About GTE Internet Solutions	http://www.gte.net/pands96.html#ISDN
UUNET for Nationwide Access	http://www.gte.net/nationwide.html
Availability and Dial-In Access Numbers	http://www.gte.net/dialin.html
Customer Support	800-927-3000 or 214-751-3800

TIP The GTE Customer Support help desk is staffed around the clock—24 hours a day, 7 days a week, 365 days per year.

Northern Arkansas Telephone Company (NATCO)

Web: http://southshore.k12.ar.us/natco1.html

Phone: 800-775-6682

E-mail: sandersj@southshore.k12.ar.us

NYNEX

Contact	Address/Phone
Home Page	http://www.nynex.com/
ISDN Information	http://www.nynex.com/isdn/isdn.html
Authorized Manufacturers, Suppliers, Service Providers, and Sales Agents	http://www.nynex.com/isdn/9xpart.html
ISDN Pricing and Ordering	http://www.nynex.com/isdn/pricing.html
Other ISDN Inquiries	800-GET-ISDN
E-mail	notes.jducay@nynex.com

Pacific Bell

Contact	Address/Phone
Home Page	http://www.pacbell.com/
FasTrak Services (Business Data Products)	http://www.pacbell.com/Products/fastrak.htm
How to Order	http://www.pacbell.com/Products/SDS-ISDN/isd-21.htm
ISDN Service Center	800-4PB-ISDN

Contact	Address/Phone
E-mail	isdn-info@pacbell.com, mention code: Net-1
Home ISDN (Pricing and Availability)	http://www.pacbell.com/Products/SDS-ISDN/Home_ISDN/
Basic Rate ISDN (Options and Pricing)	http://www.pacbell.com/Products/SDS-ISDN/isd-22.htm

SBC Communications/ Southwestern Bell

Contact	Address/Phone
SBC Home Page	http://www.sbc.com/
Southwestern Bell Home Page	http://www.swbell.com/swbell/swbell-sm.html
ISDN Home Page	http://www.sbc.com/swbell/kc/isdn.html
ISDN Pricing Information	http://www.sbc.com/swbell/kc/isdncost.html
SWBT—Communications Solutions for Small Business	http://www.sbc.com/swbell/shortsub/comm_solutions.html#top
ISDN Availability	800-SWB-ISDN

NOTE

DigiLine Service (ISDN BRI) is currently available only in certain parts of Southwestern Bell's service area of Texas, Missouri, Kansas, Oklahoma, and Arkansas. For example, DigiLine Service is available to Southwestern Bell customers in the 913 and 316 area codes in Kansas and in the 816 and 417 area codes in Missouri.

US West

Contact	Address/Phone
Home Page	http://www.uswest.com/
ISDN	http://www.uswest.com/isdn/index.html
ISDN Availability	http://www.w3.uswest.com/isdn/availability.html
ISDN Pricing	http://www.w3.uswest.com/isdn/pricing.html
How to Order	http://www.w3.uswest.com/isdn/order.html
Residence & Home Office	800-898-9675
Small Business (less than 20 exchange lines)	800-246-5226
Large Business (more than 20 exchange lines)	By state
AZ, ID, MT, OR, UT, WA, WY	800-839-4616

Contact	Address/Phone
CO	303-787-1000
IA	515-286-5139
MN, ND	612-399-7575
NB, SD	800-228-3444
NM	505-245-5600

International Carriers

We've organized the international carriers as follows:

- Canada
- Europe
- Other countries (Korea, Middle East, Australia, Japan, Africa)

Canada

NOTE Stentor is the alliance of Canada's telephone companies, who are the leading providers of telecommunications to millions of customers in Canada and around the world: AGT, BC TEL, Bell, Island Tel, MT&T, MTS, NBTel, NewTel Communications, NorthwesTel, Québec-Téléphone, and SaskTel.

Contact	Address/Phone
Stentor	http://www.stentor.ca/
Canadian ISDN Resource Centre	http://www.canisdn.net/
Bell Z@P ISDN Service	http://www.bell.ca/promo/zap/
MT&Ts Microlink Home ISDN service	http://www.canisdn.net/mtt/ micro.html
CanISDN Speed Test	http://www.canisdn.net/size.html
ISDN Availability	http://www.canisdn.net/cgi-win/ stentor/isdn.exe/eng

Europe

Contact	Address/Phone
Belgacom (Belgium)	http://www.belgacom.be/index.htm
British Telecom	http://www.bt.net/
Products and Services	http://www.bt.com/home/products/ index.htm
ISDN Helpline	0800-181-514
Deutsche Telekom	http://www.dtag.de/dtag/ telekom_.html
Products and Services	http://www.dtag.de/dtag/prodserv/ index_.html

Contact	Address/Phone
Euro-ISDN Informationen der Telekom	http://www.netcs.com/NetCS-Public/edss1.ger.html
France Telecom	http://www.francetelecom.com/
Products and Services	http://www.francetelecom.com/ft/product.htm
Domestic ISDN Tariffs (Numeris)	http://www.francetelecom.com/ft/product/numeris/numeris.htm
International ISDN Service	http://www.francetelecom.com/ft/product/isdn/desc.htm
Helsinki Telephone Company (Finland)	http://www.hpy.fi/english/index.html
Residential Products and Services	http://www.hpy.fi/english/products/res.html
Business Products and Services	http://www.hpy.fi/english/products/bus.html
Fax	+358 0 664 480
Swiss Telecom (Switzerland)	http://www.vptt.ch/
Telecom Services	http://www.vptt.ch/services.html
SwissNet (ISDN)	http://www.vptt.ch/swissnet.html
Tele Danmark	http://www.teledanmark.dk/english/english.htm

Contact	Address/Phone
Telephony Services and Pricing	http://www.teledanmark.dk/english/intro/inv_rela/information/inf5.html
Telecom Finland	http://www.inet.fi/telecom/english/
Phone	+358 20 401
Fax	+358 20 403 2032
iNetPro (ISDN)	http://www.inet.fi/inet/pro.htm#ISDN
Telecom Italia	http://www.telecomitalia.interbusiness.it/Telecom/en/index.html
Business Clients	http://www.telecomitalia.interbusiness.it/Telecom/en/imprese.html

Other Countries

Contact	Address/Phone
Korea	http://soback.kornet.nm.kr/~ktsi1/html/es-isdn0.htm
BEZEQ The Israel Telecommunication Corp. Ltd (Middle East)	.http://www.bezeq.co.il/eindex.html, http://www.bezeq.co.il/eitalking.html
ISDN Development in Israel	http://wwwold.technion.ac.il/teach/topnet/cnpp95/ISDN_in_Israel/isdn.html

Contact	Address/Phone
Telstra (Australia)	http://www.telstra.com.au/corp.html
Telstra ISDN Products and Services	http://www.telstra.com.au/prod-ser/isdn/index.htm
Nippon Telegraph and Telephone (NT&T) Corp. (Japan)	http://www.ntt.jp/index.html
NT&T ISDN Home Page	http://www.info.hqs.cae.ntt.jp/SER/ISDN/ISDN.html
Telkom S.A. (Africa)	http://www.telkom.co.za/
Telkom Marketing/Product Sales	http://www.telkom.co.za/contact.html

APPENDIX

B

ISDN Equipment Vendors

Vendor	Web Address
3Com	http://www.3com.com/
ACC	http://www.acc.com/
ADAK	http://www.adak.com/
Adtran	http://www.adtran.com/
Adtran, ISDN Extension Products	http://www.adtran.com/cpe/isdn/extnsion.html
Advanced Computer Communications	http://www.acc.com/
Ascend Communications, Inc.	http://www.ascend.com
AT&T ISDN Products	http://www.attns.com
Audio Processing Technology (APT)	http://www.aptx.com
Bay Networks	http://www.baynetworks.com/
BinTec Computersysteme	http://www.bintec.de/
Biodata	http://www.biodata.de/
Chase Research	http://www.chaser.co.uk/
Cisco Technologies	http://www.cisco.com/
Cisco Technologies: ISDN	http://cio.cisco.com/warp/public/2/ISDN.html
Connectware	http://www.connectware.com/
Controlware Communications Systems	http://www.cware.de/
Controlware, idb64_2 Link Backup Controller	http://www.cware.de/idb64_2.htm
Controlware, Taxi PRI Terminal Adapter	http://www.cware.de/taxi.htm

Vendor	Web Address
Cray Communications	http://www.craycom.com/
DATAX, BRI and PRI Backups For Frame Relay Links	http://www.telindus.be/htdocs/datax/datax.htm
DigiBoard	http://www.digibd.com/
DigiBoard: ISDN.	http://www.digibd.com/support/tips/isdn/cpe.html
Digital Equipment Corp.	http://www.dec.com/
Dynatech Communications	http://www.dynatech.com/
Eicon Technology	http://www.eicon.com/
Ericsson (U.S. mirror site)	http://www.ericsson.com/
Farallon	http://www.farallon.com
Flowpoint	http://www.flowpoint.com/
Fujitsu	http://www.fujitsu.com
Fujitsu Products & Contacts	http://www.fujitsu.com/prod.html
Gandalf Technologies	http://www.gandalf.ca/
Hadax Electronics.	http://www.linnet.com/~hadax/
Intel	http://www.intel.com/
ISC, SecureLink Server	http://www.infoanalytic.com/isc/securelk.htm
ISDN Telephone Information	http://www.deepeddy.com/~cwg/isdn/
ISDN*tek	http://isdntek.com/
Italtel	http://www.italtel.it/vcbuctbpm/buct-bpm/isdn/isdn.html
Jetstream	http://www.jetstream.com/
Lightning Instrumentation	http://www.lightning.ch/

Vendor	Web Address
Livingston Enterprises	http://www.livingston.com/
Microtronix Datacom	http://www.microtronix.com/mdlprod/
Microsoft	http://microsoft.com
Motorola	http://www.mot.com/
Motorola Information Systems Group	http://www.mot.com/MIMS/ISG/Products/
NEC	http://www.nec.com/
Network Dynamics	http://www.nd.co.nz/
Network Express	http://www.nei.com or http://branch.com/netexpress/netexpress.html
Newbridge Microsystems	http://www.newbridge.com/
Northern Telecom (Nortel)	http://www.nortel.com/
OfficePoint	http://server1.service.com/info/product/switching/office-point.html
Penril Datability Networks	http://www.penril.com/index.html
Penril, DSX 2622 ISDN Terminal Adapter	http://www.penril.com/dsx2622.html
Primary Rate Incorporated	http://www.cygnus.nb.ca/pri.html
Protean	http://protean.com
Protocol Converters	http://www.sec3net.securicor.co.uk/product/interchange/interchange.html
Racal Datacom	http://www.racal.com/
Securicor 3net	http://www.sec3net.securicor.co.uk/
Shiva Corporation	http://www.shiva.com/
Siemens Rolm Communications	http://server1.service.com/info/

Vendor	Web Address
Siemens Stromberg-Carlson	http://www.ssc.siemens.com/
Silicon Graphics	http://www.sgi.com/
Skyline Technology	http://www.skylinetech.com/
Stollmann E+V GmbH (German)	http://www.stollmann.de/
Sun Microsystems	http://www.sun.com/
Symplex Communications	http://www.symplex.com/
Telebit	http://www.telebit.com/
Tone Commander	http://www.halcyon.com/tcs/
Tone Commander, 40d 120	http://www.halcyon.com/tcs/40d120.html
TTSI / Bosch	http://www.tts.com/
US Robotics	http://www.usr.com/
Xyplex	http://www.xyplex.com/
ZyXEL	http://www.zyxel.com/

APPENDIX

C

ISDN Information Resources

This appendix lists resources for ISDN information, other than carriers (Appendix A) and equipment vendors (Appendix B). The resources are organized as follows:

- General information
- User groups in the U.S.
- User groups overseas
- Videoconferencing information
- Telecommuting information

General ISDN Information

Name	Web Address
Bellcore ISDN Home Page	http://www.bellcore.com/ISDN/ISDN.html
Computer and Communication Link Page	http://www.cmpcmm.com/cc/
Dan Kegel's ISDN Page	http://www.alumni.caltech.edu/~dank/isdn/
Datacomm-US	http://www.datacomm-us.com/
Jeff Frohwein's ISDN Page	http://fly.HiWAAY.net/~jfrohwei/isdn/
Gatech's ISDN Project	http://www.gcatt.gatech.edu/projects/isdn/isdn.html
MSGNet's ISDN Info and Hardware Reviews	http://www.msg.net/ISDN/
Sean's ISDN-O-Rama	http://www.teleconnect.com/isdn.htm

Name	Web Address
Sven's ISDN Information Base	http://igwe.vub.ac.be/~svendk/
UIUCnet@Home Information and Guidelines	http://tampico.cso.uiuc.edu/~kline/ISDN-stuff/isdn.html

ISDN User Groups in the U.S.

Name	Web Address
California ISDN Users Group	http://www.ciug.org/
Florida ISDN Users Group	http://www.ccg4isdn.com/isdn/fiug.html
New York ISDN Users Group	http://www.interport.net/~digital/index.html
North American ISDN Users Forum (NIUF)	http://www.niuf.nist.gov/misc/niuf.html
Open Communication Networks (OCN)	http://www.ocn.com/ocn/niuf/niuf_top.html
PRIDUF (Pacific Region ISDN/Data User Forum)	http://www.ptc.org/PRIDUF.html
Texas ISDN Users Group (TIUG)	http://www.crimson.com/isdn/

Overseas ISDN User Groups

Name	Web Address
Indonesian ISDN & Internet User Forum (IIIUF)	http://www.idola.net.id/i3uf/i3uf.htm
Southern African ISDN Forum (SAIF)	http://www.saif.org.za/

General Videoconferencing Information

Name	Web Address
Desktop Video Conferencing	http://fiddle.ee.vt.edu/succeed/videoconf.html
International Multimedia Teleconferencing Consortium	http://www.csn.net/imtc/
PictureTel Users Mailing List	http://www.idesign.com.au/vc/ptel-users/

Telecommuting Information

There's quite a bit of telecommuting information out there. To help you find what you need, we've categorized these resources as follows:

- Assistance for organizations and business
- Management and consultants
- Programs and initiatives
- Other resources

Telecommuting Assistance for Organizations and Businesses

Name	Web Address
Center for the Advanced Study of Telecommunications	http://128.146.105.96/cast/default.html
Electronic Commerce Resource Center (ECRC)	http://www.ecrc.ctc.com

Name	Web Address
Electronic Frontier Foundation (EFF) Telecommuting Archive	http://www.eff.org/pub/GII_NII/Telecommuting/
European Community Telework/Telematics	http://www.agora.stm.it/ectf/ectfhome.html
FIND/SVP	http://etrg.findsvp.com/index.html
Gordon & Associates	http://www.gilgordon.com/
Home Business Solutions	http://netmar.com/mall/shops/solution
Institute for the Study of Distributed Work	http://www.dnai.com/~isdw/
JALA International, Inc.	http://www.well.com/user/jala/
KLR Consulting	http://www.klr.com/klr/
Morning Star Technologies	http://www.morningstar.com/
Small Office/Home Office (SOHO) Central	http://www.hoaa.com
Smart Valley's Resource Page	http://www.svi.org/PROJECTS/TCOMMUTE/webguide
Telecommuting Advisory Council	http://www.telecommute.org
Weidenhammer Systems	http://www.hammer.net

Telecommuting Management and Consultants

Name	Web Address
4GL Corporation	http://www.4gl.com/
Telecommuting Jobs	http://www.tjobs.com
Telework Training International	http://ttihome.com/tti

Telecommuting Programs and Initiatives

Name	Web Address
Bay Area Telecommuting Assistance Project	http://www.abag.ca.gov/bayarea/telecomm/telecomm.htm
European Telework Online	http://www.eto.org.uk
Telecommute America!	http://www.att.com/Telecommute_America/
Teleprompt: The Telework Tele-Training Trial at the Institute for Computer-Based Learning	http://www.icbl.hw.ac.uk/telep/telework/conts.html
The Impact of Telecom on Travel Habits	http://www.engr.ucdavis.edu/~its/telecom/

Other Telecommuting Resources

Name	Web Address
Beyond Telecommuting: A New Paradigm for the Effect of Telecommunications on Travel	http://www.lbl.gov/ICSD/Niles/
Digital Nation	http://130.80.26.2/DN/slb.html
Home Business Review	http://www.tab.com:80/Home.Business
Ki-Net's New Organizational Structures for Engineering Design	http://www.ki-net.co.uk/ki-net/content.html
Management Technology Associates' Telework Commentary and Reports	http://www.mtanet.co.uk/mta_oen/tw_intro.htm
On Telecommuting, PS Enterprises	http://www.well.com/user/pse/telecom.htm
Pacific Bell Telecommuting Resources Guide	http://www.pacbell.com/Lib/TCGuide/index.html

Name	Web Address
Smart Valley Telecommuting Guide	http://smartone.svi.org:80/Projects/TCommute/TCGuide/
University of Tokyo Telework Report	http://www.mpt.go.jp/MPT-News/vol6-4/news4-2.html#Telework

GLOSSARY

Glossary of ISDN and Telecommunications Terms

SYMBOL

23B+D

See *Primary Rate Interface*.

2B+D

See *Basic Rate Interface*.

2B1Q

The standard line code used in ISDN service.

5ESS

A common central office switch, manufactured by AT&T, used to route ISDN calls in the phone companies' networks. ESS stands for Electronic Switching System. Currently, AT&T is divesting itself. In the future, the switch will be known as the Lucent 5ESS.

A

Acceptable Use Policy (AUP)

Refers to policies that restrict the way in which a network may be used. Usually, a network administrator makes and enforces decisions dealing with acceptable use.

Address

See *IP Address; E-mail Address*.

Address Mask

Used to identify the parts of an IP address that correspond to the different sections (separated by dots). It's also known as the *subnet mask* since the network portion of an address can be determined by the encoding inherent in an IP address.

ADSL (Asymmetrical Digital Subscriber Line)

A telecommunications technology that is currently being developed to provide a robust, full-duplex, reliable, and dedicated access digital service. It offers three channels on the same copper pair(s) of wire connection: a high-speed channel (downstream), a slower duplex (upstream) channel, and a third channel for analog phone service. *Asymmetric* refers to the fact that the downstream half of the duplex (from the central office to home) accommodates 6 Mbps; upstream is only 640 kbps.

Algorithm

A well-defined rule or process for arriving at a solution to a problem. In videoconferencing, this refers to the compression software technique that compresses the video signal.

Amplitude

The maximum value of an analog or digital waveform.

Analog Data

A physical representation of information that bears an exact relationship to the original information in the form of a wave. For example, the electrical signals on a regular telephone channel are the analog data representation of the original voice volume and frequency.

Anonymous FTP

By using the word *anonymous* as your user ID and your e-mail address as the password when you log in to an FTP site, you can bypass local security checks and gain limited access to public files on the remote computer. This type of access is available on most FTP sites, but not all.

API (Application Programming Interface)

A specification of function call conventions that defines an interface to a service.

Application Layer

See *OSI Reference Model*.

Application Sharing

In videoconferencing and dataconferencing, a feature that allows two people to work together when one of the individuals doesn't have the same application or the same version of the application. In application sharing, one user launches the application, and it appears to run on both users' machines simultaneously, although it actually is running on only one. Both users can input information and otherwise control the application using the keyboard and mouse. Files associated with the application can be easily transferred, so the results of the collaboration are available to both users immediately. The person who launched the application can lock out the other person from making changes, so the locked-out person sees the application running but cannot control it.

Archie

A way of automatically gathering, indexing, and sometimes even retrieving files on the Internet. You'll usually hear the phrase "Archie search."

Archive

A collection of files stored on an Internet machine. FTP sites are known as archives.

Asymmetrical

Refers to an architecture or network that is hierarchical, or "one to many," in nature. For example, VSAT (very small aperature terminal) is referred to as asymmetrical in that there is one source broadcasting to many various sites. Also called *half-duplex*.

Asynchronous Time Division Multiplexing (ATDM)

A method of sending information in which normal time division multiplexing (TDM) is used, except that time slots are allocated as needed rather than preassigned to specific transmitters.

Asynchronous Transfer Mode (ATM)

A data-transfer method that dynamically allocates bandwidth using a fixed-size packet, or cell. ATM, also known as *fast packet* or *cell relay*, is a high-speed networking technology capable of concurrently carrying voice, video, data, and facsimile. With current lease-line technology, such as in T1 networks, data and voice can share the link, but they are separated into different channels within the T1 stream. ATM, by contrast, intermixes traffic freely, based on the dynamic demands of the network users. LAN traffic, for instance, tends to be bursty in nature. During a LAN burst, the ATM network can dynamically provide additional bandwidth to carry the load. When the burst ends, the bandwidth is available to other users.

ATM uses low-overhead, fixed-length, 53-byte cells. Because the cells are a small fixed length, high-speed hardware switches can be built to allow data speeds from 1.5 Mbps to 1.2 Gbps. ATM will provide the basis for future Broadband ISDN (B-ISDN) standards.

Asynchronous Transmission

Operation of a network system wherein events occur without precise clocking. In such systems, individual characters are usually encapsulated in control bits called start and stop bits, which designate the beginning and ending of characters.

AT&T 5ESS

See *5ESS*.

ATDM

See *Asynchronous Time Division Multiplexing*.

ATM

See *Asynchronous Transfer Mode*.

Attenuation

Loss of communication signal energy.

AUP

See *Acceptable Use Policy*.

Authentication

Any network security process that ensures that users are who they say they are when they log in. When you type your name and password, you are authenticated and allowed access.

B

B Channel

The bearer channel of an ISDN line, which is a full-duplex, 64-kbps channel employed to send user data. The B channel transports the data, images, or any other information at either 56 or 64 kbps.

Back Channel

A channel used for sending data in the opposite direction as the primary channel. Back channels are frequently used to send control information. Using back channels, information can be delivered even if the primary channel is malfunctioning.

Back End

A node or software program that provides services to a front end.

Backbone

A network acting as a primary conduit for traffic that is often both sent from and destined for other networks.

BACP

See *Bandwidth Allocation Control Protocol*.

Bandwidth

Precisely, this refers to the difference between the highest and lowest frequencies of a transmission, measured in hertz. Most people loosely refer to bandwidth as the amount of data that can be transferred over a network connection or to describe the rated throughput capacity of a given network medium or protocol.

Bandwidth Allocation Control Protocol (BACP)

The set of rules that manage bandwidth over PPP dynamic multilink connections.

Bandwidth Reservation

In circuit-switched lines, a feature in which call bandwidth can be reserved for high-bandwidth or high-priority calls.

BARRNet

See *Bay Area Regional Research Network.*

Baseband

A characteristic of a network technology where only one carrier frequency is used. Baseband can be contrasted with *broadband*, where multiple carrier frequencies are used. Ethernet is an example of a baseband network.

Basic Rate Interface (BRI)

The ISDN interface composed of two B channels and one D channel for circuit-switched communication of voice, data, and video. BRI, also commonly known as 2B+D, is the standard ISDN line used for residential or single-user configurations and one of two subscriber interfaces. BRI contains two 64-kbps B channels and one 16-kbps D channel. The two B channels are used for voice, video, or high-speed data. The D channel is used for signaling and low-speed data transmission.

Baud

A unit of signaling speed equal to the number of discrete conditions or signal events per second. Baud is synonymous with bits per second if each signal event represents exactly one bit.

Bay Area Regional Research Network (BARRNet)

A network serving the San Francisco Bay Area. BARRNet's backbone is composed of four University of California campuses (Davis, Berkeley, Santa Cruz, and San Francisco), Stanford University, Lawrence Livermore National Laboratory, and NASA Ames Research Center.

BBS

See *Bulletin Board System.*

Bearer Channel

A bearer channel is a channel which carries data, voice, images, or any other information. In an ISDN line, the B channels are the bearer channels.

Bearer Services

The options of CSV, CSD, and PSD are broad categories of bearer services that the phone companies can provide. Different bearer services provide different types of guarantees about the reliability and synchronization of the data. Bearer services are defined in terms of a number of attributes, which include mode (circuit or packet), structure (bitstream or octet-stream), transfer rate (e.g., 64 kbps), transfer capability (basically, the content, for instance speech, 7 KHz audio, video, or unrestricted), and several other attributes that specify protocols to use and other items. The attributes of the bearer service are encoded into a Bearer Code (BC) that is sent every time a new connection is being set up.

Bell Operating Company (BOC)

One of the local telephone companies that existed (prior to deregulation, under which AT&T was ordered by the courts to divest itself) in each of the seven United States regions.

Bellcore (Bell Research Corporation)

An organization that performs research and development on behalf of the seven RBOCs.

BERT (Bit Error Rate Tester)

A device that determines the bit error rate on a given communications channel.

B-ISDN

See *Broadband ISDN.*

Bit Rate

The speed at which bits are transmitted, usually expressed in bits per second (bps).

BITNET

Because It's Time Network. A low-cost, low-speed academic network consisting primarily of IBM mainframes and 9600-bps leased lines. Remote job entry (RJE) is the primary means for performing work on this network, which recently merged with CSNET (Computer and Science Network) to form CREN (Corporation for Research and Educational Networking). A more recent version of BITNET (known as BITNET-II) encapsulates the BITNET protocol within IP packets.

Black Hole

A routing term for an area of the internetwork where packets enter but do not emerge, due to adverse conditions or poor system configuration within a portion of the network.

Blocking

In a switching system, a condition in which no paths are available to complete a circuit. The term is also used to describe a situation in which one activity cannot begin until another has been completed.

BOC

See *Bell Operating Company*.

BONDING

For Bandwidth ON Demand InterNetworking Group, the method of combining two 64-kbps B channels together to increase bandwidth capacity to 128 kbps. This is done by establishing two B channel circuit-switched calls and combining them by using the Multilink PPP or BONDING specification or by associating multiple B channels

in the network (multirate services). BONDING is often used with the BRI ISDN configuration.

BRI

See *Basic Rate Interface*.

Bridge

A device that connects two or more physical networks and forwards packets between them.

In IEEE 802 parlance, a bridge is a device that interconnects LANs or LAN segments at the Data-Link layer of the OSI model to extend the LAN environment physically. LAN bridges work with frames (as opposed to packets) of data, forwarding them between networks.

In videoconferencing terminology, a bridge connects three or more conference sites so that they can simultaneously communicate. Bridges are often called MCUs, for multipoint conferencing units.

In audioconferencing, the term bridge is used to refer to a device that connects multiple (more than two) voice calls so that all participants can hear and be heard.

Broadband

A transmission system that multiplexes multiple independent signals onto one cable. Broadband can be contrasted with *baseband*, which is a characteristic of a network technology where only one carrier frequency is used.

In telecommunications terminology, any channel having a bandwidth greater than a voice-grade channel (4 KHz). In LAN terminology, a coaxial cable on which analog signaling is used. Also called *wideband*.

Broadband ISDN (B-ISDN)

Communication standards being developed by the IT-U to handle high-bandwidth applications such as video. B-ISDN will use ATM technology over SONET-based transmission circuits to provide data rates of 155 Mbps to 622 Mbps and beyond.

Browser

A software package that can be used to search the World Wide Web, display Web documents, and allow the user to select hyperlinks to other Web documents. The most popular browsers, such as Mosaic and Netscape, use graphical software and run on top of graphical user interfaces (GUIs).

Bulletin Board System (BBS)

A computer which typically provides e-mail services, file archives, and announcements of interest to the bulletin board system's operator (known as a sysop). BBSs started out as hobbies for computer enthusiasts, and were mostly accessible by modem. Recently, however, more and more BBSs are being connected to the Internet.

Bus Topology

Linear LAN architecture in which transmissions from network stations propagate the length of the medium and are received by all other stations. (Analogous to a city bus line with each bus stop representing a network station.)

C

CO

See *Central Office*.

Cable Modem

A new data service utilizing coaxial cable. This is also the customer terminal equipment required for the service.

CALC

See *Customer Access Line Charge*.

Call Bumping

A process where a second data link is dropped for an incoming voice call.

Call Priority

Priority assigned to each origination port in circuit-switched systems. This priority defines the order in which calls are reconnected. Call priority also defines which calls can or cannot be placed during a bandwidth reservation.

Call Setup Time

The time required to establish a switched call between data terminal equipment (DTE) devices.

Caller ID

A phone service feature that allows the called party to know the numbers from which calls originate. Also called calling line identification (CLID). With this feature, various call-management tools are available, such as call blocking.

Carrier Signal

A signal suitable for modulation by another signal containing information to be transmitted.

CATV (Cable Television)

Formerly called Community Antenna Television, a communication system where multiple channels of programming material are transmitted to homes using broadband coaxial cable.

CBDS

See *Connectionless Broadband Data Service.*

CCITT (Consultative Committee International Telegraph and Telephone)

An organization established by the United Nations to develop worldwide standards for data communications, now renamed ITU-T, for its parent organization International Telecommunications Union, and its sector, Telecommunications Standards. Every four years, the organization updates the standards; the most recent was 1996.

CCS (Common Channel Signaling)

A signaling system used by many telephone networks that separates signaling information from user data. Exclusive use of a specified channel to carry signaling information for all other channels in the group.

Cell

The basic unit for ATM switching and multiplexing. Each cell consists of a 5-byte header and 48 bytes of payload.

Cell Relay

Network technology based on the use of small, fixed-size packets, or cells. Cells contain identifiers that specify the data stream to which they belong. Because the cells are fixed length, they can be processed and switched in hardware at very high speeds. Cell relay is the basis for many high-speed network protocols, including IEEE 802.6, DQDB, the SMDS Interface Protocol, and ATM.

Central Office (CO)

The telephone company's local facility that provides telephone service in your area. The office connects all local loops within a given area and performs the circuit switching for the subscriber lines.

Centrex

A service provided by a central office that provides a virtual PBX to a set of extensions. It offers features such as call transfer, conference, and forward within that set of extensions. An improved PBX, Centrex also provides direct inward dialing and automatic number identification of the calling PBX. This term is also used to refer to a specific AT&T telephone system product.

CEPT (Conference Europeane des Postes et Telecommunications)

An association of the 26 European PTTs (postal telephone and telegraph companies) that recommends communication specifications to the CCITT (now ITU-T). This group has been superseded by other European organizations.

CERFNET

For California Education and Research Foundation Network, a TCP/IP-based network in Southern California connecting many higher-education centers; designed to advance science and education through communications.

Challenge-Handshake Authentication Protocol (CHAP)

A security feature that prevents unauthorized access to devices running the feature. This authentication method is often used when connecting to an Internet service provider. CHAP allows users to log in to their provider automatically, without the need for a terminal screen. It is more secure than the Password Authentication Protocol (PAP, another widely used authentication method) since it does not send passwords in text format. CHAP is supported only on lines using PPP encapsulation.

Channel Service Unit (CSU)

A digital interface device that connects end-user equipment to the local digital telephone loop.

CHAP

See *Challenge-Handshake Authentication Protocol*.

Chat

Another term for IRC (Internet relay chat). Also, an acronym for Conversational Hypertext Access Technology.

CIF (Common Intermediate Format)

An international standard for video display formats. The QCIF format, which employs half the CIF spatial resolution in both horizontal and vertical directions, is the mandatory H.261 format. QCIF is used for most desktop videoconferencing applications where head-and-shoulder pictures are sent from desk to desk. QCIF displays 176 pixels grouped in 144 noninterlaced luminance lines.

Circuit

A communications link between two or more points.

Circuit Switching

A switching system in which a dedicated physical circuit path must exist between sender and receiver for the duration of the "call." Used heavily in the phone-company network, circuit switching often is contrasted with contention and token passing as a channel-access method, and with message switching and packet switching as a switching technique.

Clear To Send (CTS)

A circuit in the RS-232 specification that is activated when the data communications equipment (DCE) is ready to accept data from the data terminal equipment (DTE).

CLID

See *Caller ID*.

Client

In Internet terminology, an application that performs a specific function, such as Telnet or FTP. It's the "front-end" to an Internet process.

In more general terms, a client is computer system or process that requests a service of another computer system or process.

Client/Server Computing

Term used to describe distributed processing (computing) network systems in which transaction responsibilities are divided into two parts: client (front-end) and server (back-end). Both terms (client and server) can be applied to both software programs or actual computing devices.

CLNP/CLNS (Connectionless Network Protocol/Connectionless Network Service)

A synonym for SMDS, a high-speed, packet-switched, datagram-based WAN technology. This is an OSI Network-layer protocol/service that does not require a circuit to be established before data is transmitted. CLNP is the OSI equivalent of IP.

CO

See *Central Office*.

Coaxial Cable

A cable consisting of a hollow outer cylindrical conductor that surrounds a single inner wire conductor. Two types of coaxial cable are currently used for LANs: 50-ohm cable, which is used for digital signaling, and 75-ohm cable, which is used for analog signaling and high-speed digital signaling.

Codec

For COmpression/DECompression, a set of hardware or software components, or a combination, providing digital compression and decompression of analog video signals so that they can be efficiently transmitted or stored. The compression method may be proprietary or standards-based. In videoconferencing, the codec also acts as an interface. Video and audio are run through the codec, which transmits a single, digital signal over a network to the remote location(s).

Codec Conversion

The back-to-back transfer of an analog signal from one codec into another codec in order to convert from one proprietary coding scheme to another. The analog signal, instead of being displayed to a monitor, is delivered to the dissimilar codec. There, it is redigitized, compressed, and passed to the receiving end. Conversion service is offered by carriers such as AT&T, MCI, and Sprint.

Coder/Decoder

A device that typically uses pulse code modulation (PCM) to transform analog voice into a digital bit stream and vice versa.

Common Carrier

A licensed, private utility company that supplies communication services to the public at regulated prices.

Common Channel Signaling

See CCS.

Communications Server

A communications processor that connects asynchronous devices to a LAN or WAN through network and terminal emulation software. Communication servers are generally used to pool modems on a network. These pooled modems are shared among network users to make dial-out connections to the Internet, to information services (like CompuServe and Lexis-Nexis), for faxing, alpha-numeric paging, and more. Communication servers are also used for dial-in remote-control applications.

Communications Line

The physical link (such as a wire, cable, or a telephone circuit) that connects one or more devices to another.

Composite Video

A television signal where the chrominance signal is a sine wave that is modulated onto the luminance signal, which acts as a subcarrier. This is used in NTSC and PAL (United States and European video standards, respectively) systems.

Compressed Serial Link Internet Protocol

See CSLIP.

Compression

The process of reducing the information content of a signal so that it occupies less space on a transmission channel or storage device. Compression is a fundamental concept of video communications. Compression of video information can be accomplished by reducing the quality (sending fewer frames in a second or displaying the information in a smaller window) or by eliminating redundancy.

Computer Science Network
See CSNET.

Computer Telephony Integration (CTI)
The integration of telephony features into the PC, so that telephone functions can be handled in one unit. Ideally, you will need just your PC for phone, PC, voice mail, and fax functions.

Connectionless Network Protocol/Connectionless Network Service
See *CLNP/CLNS*.

CONS (Connection-Oriented Network Service)
An OSI protocol providing connection-oriented operation to upper-layer protocols.

Contention
An access method in which network devices compete for the right to access the physical medium.

Convergence
The ability of (and speed with which) a group of internetworking devices, running a specific routing protocol, agree on the internetwork's topology after a change in network topology.

Corporation for Research and Educational Networking
See *CREN*.

COSINE
For Corporation for Open Systems Interconnection Networking in Europe, a European project financed by the European Community (EC) to build a communications network between scientific and industrial entities in Europe.

CPE
See *Customer Premises Equipment.*

CREN (Corporation for Research and Educational Networking)
An organization formed in October 1989, when BITNET and CSNET were combined. CSNET is no longer around, but CREN still operates BITNET.

Crosstalk
Interference generated when magnetic fields or other nearby circuits interrupt electrical currents in a circuit. Crosstalk results in signal distortion.

CSD (Circuit-Switched Data)
One of two provisioning features for ISDN. CSD specifies telephone calls, with data only.

CSLIP (Compressed Serial Link Internet Protocol)
A protocol that minimizes traffic and speed throughput on SLIP lines.

CSNET (Computer Science Network)
A large internetwork consisting primarily of universities, research institutions, and commercial concerns. CSNET has merged with BITNET to form CREN.

CSU
See *Channel Service Unit.*

CSV (Circuit-Switched Voice)

One of two provisioning features for ISDN. CSV is your basic and traditional telephone service.

CTI

See *Computer Telephony Integration*.

CTS

See *Clear To Send*.

Customer Access Line Charge (CALC)

A federal tariff for hooking up your ISDN line. Also referred to as an End User Common Line Charge (EUCL) or Subscriber Line Charge (SLC). Every ISDN line is charged one CALC or EUCL. The amount varies by state or province.

Customer Premises Equipment (CPE)

In telecommunications, any equipment used at the customer's location which is not owned by the telephone company. For ISDN service, CPE might include NT1s (network terminating devices), terminal adapters, and digital phones installed at customer sites and connected to the phone company network.

D

D Channel

The signaling channel of an ISDN line. In PRI, this is a 64-kbps channel; in BRI, it is 16 kbps. The D channel can also be used to carry packet-switched data.

Data Circuit-Terminating Equipment

See *Data Communications Equipment*.

Data Communications Equipment (DCE)

The devices and connections of a communications network that connect the communication circuit with the end device, which is the data terminal equipment (DTE). A modem can be considered DCE.

Data Compression

Reducing the size of a data file by removing unnecessary information, such as blanks and repeating or redundant characters or patterns.

Data-Link Layer

See *OSI Reference Model*.

Data Service Unit (DSU)

A device used in digital transmission for connecting data terminal equipment (DTE), such as a router, to a digital transmission circuit (DTC) or service.

Data Set Ready (DSR)

An RS-232-C interface circuit that is activated when the data communications equipment (DCE) is powered up and ready for use.

Data Terminal Equipment (DTE)

The equipment that serves as a data source, destination, or both. The DTE provides the controls for data communications according to protocols. DTE includes computers, protocol translators, and multiplexers. The data communications equipment (DCE) connects the communication circuit with the DTE.

Data Terminal Ready (DTR)

An RS-232-C circuit that is activated to let the data communications equipment (DCE) know when the data terminal equipment (DTE) is ready to send and receive data.

Dataconferencing

Exchanging files, information, programs, and applications through a computer conference with two or more participants.

Datagram

A "smart" block of data that carries information from one Internet site to another without requiring earlier exchanges between the source and destination computers. IP datagrams are the primary information units in the Internet. The terms *packet*, *frame*, *segment*, and *message* are also used to describe logical information groupings at various layers of the OSI reference model.

DATANET

A major Netherlands public switched telephone network.

DATAPAC

A large Canadian public switched telephone network.

DATAPAK

A packet-switched public network in the Nordic countries (Denmark, Finland, Iceland, Norway, and Sweden).

DATEX-L

A circuit-switched public network in Germany.

DATEX-P

A packet-switched public network in Germany.

DCE

See *Data Communications Equipment; Distributed Computing Environment.*

DDR

See *Dial-on-Demand Routing.*

DECnet

A group of communications products (including a protocol suite) developed and supported by Digital Equipment Corporation (DEC). Also, the proprietary network protocol designed by DEC.

Decryption

See *Encryption.*

Dedicated Line

A communications line that is used solely for computer connections. If you buy an additional phone line for your modem, that's a dedicated line. There are other types of dedicated lines (such as T3s and T1s) that are used for larger networks. When the line is not owned by the user, the term *leased line* is more common.

Demarc

Short for the demarcation point, or the point where the telephone company's wiring stops and the customer's wiring begins. The phone company charges extra for any wiring work performed on the customer's side of the demarc.

Demodulation

The process of returning a modulated signal to its original form. Modems perform demodulation by converting an analog signal to its original (digital) form.

Demultiplexer

A device that separates multiplexed material from a single input and the individual elements to multiple output streams.

Dial Backup

A feature supported by some routers that provides protection against WAN downtime. Dial backup allows the network administrator to configure a backup serial line through a circuit-switched connection.

Dial-on-Demand Routing (DDR)

A routing feature that provides on-demand network connections. The router can automatically initiate and close a circuit-switched session. DDR permits routing over ISDN or phone lines using an external ISDN terminal adapter or modem.

Dial-Up Connection

A connection that uses regular phone lines to connect one computer to another via modem. Dial-up lines are a common method of accessing the Internet.

Digital

Of or relating to the technology of computers and data communications wherein all information is encoded as ones or zeros. The ones and zeros represent on or off states, known as *bits*.

Digital Network Architecture (DNA)

Digital Equipment Corporation's network architecture. DECnet is the term used to refer to products (including communications protocols) that are based on the DNA.

Distance Learning

The incorporation of video and audio technologies into the educational process so that students can attend classes and training sessions in a location distant from where the course is being presented. Distance learning systems are usually interactive. The technology brings the classroom to the student versus the traditional arrangement of the student traveling to the classroom.

Distributed Computing Environment (DCE)

An architecture based on standard programming interfaces, conventions, and server functionality used for distributing applications transparently across networks. The DCE is controlled and promoted by the Open Software Foundation (OSF), a consortium of vendors including Digital Equipment Corporation, IBM, and Hewlett-Packard.

DMS-100

A type of central office equipment switch that is manufactured by Northern Telecom. DMS stands for Digital Multiplexing System (or Switch).

DNA

See *Digital Network Architecture*.

DNS

See *Domain Name System*.

Domain

Often referred to loosely as *site*, a logical region of the Internet, or a portion of a name hierarchy tree. Generally, a domain corresponds to an IP address or an area on a host.

Domain Name System (DNS)

A static, hierarchical name service used with TCP/IP hosts and housed on a number of servers on the Internet. The DNS maintains a database for finding (or resolving) host names and IP addresses on the Internet. This allows users to specify remote computers by host names rather than numerical IP addresses.

Domestic Trunk Interface (DTI)

A circuit used for DS-1 applications with 24 trunks.

DOMPAC

A large French Guyana public switched telephone network.

Dot Address

The common notation for IP addresses, also known as *dotted-decimal notation*, of the form 1.2.3.4, where each number represents one byte in the four-byte IP address.

Download

To retrieve a file from another computer and store it in your machine's own memory. Communications software that facilitates file transfer can be used to transfer and receive information, programs, and instructions from a central or larger computer to the local workstations or desktop computers.

Downstream Communication

In data transmission, communication from the data source. *Upstream* communication is the communication to the data source.

DS-0

A single 64-kbps channel of a DS-1 digital facility.

DS-1

Refers to Digital (transmission) System 1, or Digital Signal level 1, the 1.544-Mbps (U.S.) or 2.048-Mbps (Europe) digital signal carried on a T1 or E1 facility.

DS-3

Refers to Digital (transmission) System 3, or Digital Signal level 3 the 44.736-Mbps digital signal carried on a T3 facility.

DS-4

Refers to Digital Signal level 4, Bell System terminology for the 274.176-Mbps signal.

DSR

See *Data Set Ready*.

DSU

See *Data Service Unit*.

DSX-1

Cross-connection point for DS-1 signals.

DTE

See *Data Terminal Equipment*.

DTI

See *Domestic Trunk Interface*.

DTMF

See *Dual-Tone Multifrequency*.

DTR

See *Data Terminal Ready*.

Dual-Tone Multifrequency (DTMF)

Use of two simultaneous voiceband tones for dialing (such as touch-tone).

Dynamic Address Resolution

Use of an address resolution protocol to determine and store network address information on demand.

Dynamic Routing

Routing that adjusts automatically to network topology or traffic changes.

E

E.164

ITU (CCITT) recommendation for international telecommunication numbering, especially ISDN, B-ISDN, and SMDS. An evolution of normal telephone numbers.

E-1

A type of digital carrier system transmitting voice or data at 2.048 Mbps, primarily used in Europe.

E-3

A type of digital carrier system with a 34.368-Mbps transmission rate, the highest generally available in the European digital infrastructure.

Echo

In telephone and videoconferencing applications, reflected signals caused by impedance mismatches, or points where energy levels are not equal.

Echo Suppresser

Used to reduce annoying echoes in the audio portion of a videoconference. An echo suppresser is a voice-activated "on/off" switch that is connected to the four-wire side of a circuit.

EDI

See *Electronic Data Interchange.*

EDIFACT

See *Electronic Data Interchange for Administration, Commerce, and Transport.*

EFF

See *Electronic Frontier Foundation.*

EISI

See *European ISDN Standard Institute.*

Electromagnetic Interference (EMI)

Interference by electromagnetic signals, which can affect data integrity and increase error rates on transmission channels.

Electronic Data Interchange (EDI)

The electronic communication of operational data, such as orders and invoices between organizations.

Electronic Data Interchange For Administration, Commerce, And Transport (EDIFACT)

A data exchange standard administered by the United Nations to be a multi-industry EDI standard.

Electronic Frontier Foundation (EFF)

A foundation that addresses social and legal issues arising from the impact of computers on society.

EMA

See *Enterprise Management Architecture.*

E-Mail Address

The information required to transmit e-mail messages electronically between end users over various types of networks. An e-mail address is made up of several parts. The first part of the address, the username, identifies a unique user on a server. The "@" (at symbol) separates the username from the host name. Large servers, such as those used at universities or large companies, sometimes contain multiple parts, called subdomains. Subdomains and the host name are separated by a "." (dot). The three-letter suffix in the host name identifies the kind of organization operating the server. The most common suffixes are com (commercial), edu (educational), gov (government), mil (military), net (networking), and org (noncommercial). Addresses outside the United States sometimes use a two-letter suffix that identifies the country in which the server is located. A few examples are jp (Japan), nl (Netherlands), uk (United Kingdom), ca (Canada), and tw (Taiwan).

EMI

See *Electromagnetic Interference.*

Encryption

The basis of network security. Encryption encodes network packets to prevent anyone except the intended recipient from accessing the data. In encryption, a specific algorithm is applied to the data to alter its appearance and make it incomprehensible to those who might attempt to "steal" the information. The process of *decryption* applies the algorithm in reverse to restore the data to its original appearance.

End of Transmission (EOT)

Generally, a character that signifies the end of a logical group of characters or bits.

End System

Generally, an end-user device on a network. Also, a nonrouting host or node in an OSI network.

Enterprise Management Architecture (EMA)

Digital Equipment Corporation's network management architecture, based on the OSI network management model.

Enterprise Network

A large, diverse network connecting most major points in a company. An enterprise network differs from WAN in that it is typically private and contained within a single organization.

EOT

See *End of Transmission.*

Equalization

A technique used to compensate for communications channel distortions.

ESNET

Energy Sciences Network, a multinational internetwork.

ETNO (European Telecommunications Network Operators)

A European association of network providers and carriers in Europe.

ETSI (European Telecommunications Standards Institute)

A European standards body managing the implementation of Euro-ISDN.

Ethernet

A baseband LAN specification standard that is a popular connection type for LANs. In an Ethernet configuration, computers are connected by coaxial or twisted-pair cable where they contend for network access using a carrier sense multiple access with collision detection (CSMA/CD) method. Ethernet can transfer information at up to 10 Mbps. Ethernet is similar to a series of standards produced by IEEE referred to as IEEE 802.3.

Ethertalk

Appletalk protocols running on Ethernet.

ETSI

See *European Telecommunications Standards Institute*.

EUCL (End User Customer Line)

See *Customer Access Line Charge*.

EUNET

European Unix network designed to provide interconnection and e-mail services.

Euro-ISDN

An ETSI standard for ISDN.

Euronet

A networking scheme proposed by European common market countries.

European Computer Manufacturers Association

A group of European computer vendors that have done substantial OSI standardization work.

European ISDN Standard Institute (EISI)

An organization created by the European PTTs (postal telephone and telegraph companies) and the EC (European Community) to propose telecommunications standards for Europe.

European Telecommunications Standards Institute

See *ETSI*.

EWSD

Central office equipment manufactured by Siemens Stromberg-Carlson.

Exchange

See *Central Office*.

Expansion

The process of running a compressed data set through an algorithm that restores the data set to its original size.

F

FAQ (Frequently Asked Question)

Widely available on the Internet, usually in the form of large, instructional text files. FAQs are written on a large variety of topics, and they are usually the most up-to-date source for specialized information.

Fast Switching

A router feature whereby a cache is used to expedite packet switching through a router.

FCC (Federal Communications Commission)

A government agency that supervises, licenses, and controls electronic and electromagnetic transmission standards.

FDDI (Fiber Distributed Data Interface)

Set of ANSI/ISO standards that define a high-speed (100-Mbps) LAN standard using fiber-optic cabling as the transfer medium. FDDI is a powerful improvement of the token-ring network, with a maximum length of 60 to 120 miles and a total of 500 to 1000 stations. The two-core optical fiber provides two rings, a primary ring and a secondary ring; the secondary ring is operated as a backup. FDDI II is the proposed ANSI standard to enhance FDDI by providing isochronous transmission for connectionless data circuits and connection-oriented voice and video circuits.

FDM

See *Frequency Division Multiplexing.*

Federal Networking Council (FNC)

A collection of federal agencies that have heavy interests in federal networks using TCP/IP and the Internet. Representatives from DOD, DOE, DARPA, NSF, NASA, and HHS are the major members of the FNC.

Fiber Distributed Data Interface

See *FDDI.*

Fiber-Optic Cable

A thin, flexible medium capable of conducting modulated light transmission. Compared with other transmission media, fiber-optic cable is more expensive, immune to electromagnetic interference, and capable of higher data rates.

File Transfer

One of the most popular network applications, whereby files can be moved from one network device to another.

File Transfer Protocol

See *FTP.*

Finger

A Unix command that shows information about a user or group of users on the Internet. When executed, the Finger command usually returns the user's real name, whether or not the user has unread mail, and the time and date of that user's last login. Finger also displays two files (if they exist) located in the home directory of the user fingered. These two files (the .PLAN and .PROJECT files.) are simply ASCII text files that can be opened by the user to display any information upon being fingered.

Flapping

A routing problem where the advertised route between two nodes alternates (flaps) back and forth between two paths. Flapping is caused by a network problem that causes intermittent interface failures.

Flooding

A routing technique by which routing information received by a routing device is sent out to each of that device's interfaces except (usually) the interface on which the information was received.

FNC

See *Federal Networking Council.*

Forward Channel

The communications path carrying information from the call initiator to the called party.

Fragmentation

The process of breaking a packet into smaller units when transmitting over a network medium that cannot support the original size of the packet.

Frame

A logical grouping of information sent as a Data-Link-layer unit over a transmission medium. The terms *packet*, *datagram*, *segment*, and *message* are also used to describe logical information groupings at various layers of the OSI reference model.

Frame Relay

A fast packet-switching technology that offers switched virtual transmission services at speeds up to T1. A protocol used across the interface between user devices (for example, hosts and routers) and network equipment (for example, switching nodes). Frame relay is a more efficient replacement for X.25.

Frame Store

A system capable of storing complete frames of video information in digital form. This system is used for television standards conversion, computer applications incorporating graphics, video walls (a series of four or more monitors, stacked to form a square, each coordinated by a computer to show a segment of a picture, video, or film), and various video production and editing systems.

Freenet

A network system made up of community-based bulletin board systems with e-mail, information services, interactive communications, and conferencing. Freenets are usually funded and operated by individuals or organizations much like public television. Freenet providers are part of the National Public Telecomputing Network (NPTN), a Cleveland-based organization that works to make computer networking services as freely available as public libraries.

Frequency

Measured in hertz (Hz), the number of cycles of an alternating current signal per a unit of time.

Frequency Division Multiplexing (FDM)

A technique whereby information from multiple channels can be allocated bandwidth on a single wire based on frequency.

Frequently Asked Questions

See *FAQ*.

FTP (File Transfer Protocol)

The most widely used way of downloading and uploading files across an Internet connection. FTP is a standardized way to connect computers so that files can be shared between them easily. Formerly, all FTP connections were text-based, but graphical applications are now available that make FTP commands as easy as dragging and dropping. Numerous FTP clients exist for a number of platforms.

Full-Duplex (FDX)

Two-way, simultaneous transmission of data; a communication protocol in which the communications channel can send and receive data at the same time. Compare to half-duplex, where information can only be sent in one direction at a time. The telephone network is full-duplex.

Full Common Intermediate Format (FCIF)

See *CIF*.

Full-Motion Video

Video reproduction at 30 frames per second (fps) for NTSC signals, or 25 fps for PAL signals. Also known as continuous-motion video. In the video-conferencing world, equipment vendors often use "full-motion video" to refer to any system that isn't still-frame. Most videoconferencing systems today run 10 to 15 fps at 112 kbps.

G

G.703

An ITU electrical and mechanical specification for connection between phone company digital equipment and data terminal equipment (DTE).

Gateway

A "mediator" device or program that passes information between networks that normally couldn't communicate. In the IP community, gateway is an older term referring to a routing device. Today, the term *router* is used to describe nodes that perform this function.

Gopher

An information search and retrieval tool used widely for research. Gopher information is stored hierarchically on computers across the Internet. It uses a simple protocol that allows a client to access information from a multitude of numerous Gopher servers at one time, creating what's known as Gopherspace. The most common search tools in

Gopher are Veronica and Jughead. Gopher clients exist for most platforms.

GOSIP (Government OSI Profile)

A government procurement specification for OSI protocols. Through GOSIP, the government has mandated that all federal agencies standardize on OSI and implement OSI-based systems as they become commercially available.

Grade of Service

Measure of telephone service quality based on the probability that a call will encounter a busy signal during the busiest hour of the day.

H

H Channel

A full-duplex Primary Rate Interface ISDN channel operating at 384 kbps.

H.320

A recommendation of the ITU-T for video compression. H/320 can be a video system's sole compression method or a supplementary algorithm, used instead of a proprietary algorithm when two dissimilar codecs need to interoperate. H.320 includes a number of individual recommendations for coding, framing, signaling, and establishing connections. It also includes three audio algorithms: G.721, G.722, and G.728.

Half-Duplex (HDX)

The capability for data transmission in only one direction at a time. A half-duplex system is capable of transmitting data alternately, but not simultaneously, in two directions. Two communicating nodes may therefore operate as receiver and transmitter alternately. This is in contrast to full-duplex, which allows two-way, simultaneous transmission of data.

Half Gateway

Literally, a device that performs the functions of half of a gateway. Gateways are often divided into two functional halves to simplify design and maintenance.

Handset

The part of a telephone containing the transmitter and receiver that is handled during use.

Handshake

The electrical exchange of predetermined signals by devices wishing to set up a connection. For example, handshaking is used in video communications by codecs to agree on a common algorithm. Once the handshaking process is completed, the transmission begins.

Hardware Address

Also called *physical address* or *MAC-layer address*, a Data-Link layer address associated with a particular network device. A hardware address is in contrast with a network or protocol address, which is a Network layer address.

HDSL (High-Bit Rate Digital Subscriber Line)

Technology using adaptive signal processing to create two-way T1 capacity on a normal, copper twisted-pair wire without using repeaters. Also sometimes called HSDL, for High-Speed Subscriber Data Line.

Hertz (Hz)

A measure of frequency or bandwidth, synonymous with cycles per second.

Heterogeneous Network

A network consisting of dissimilar devices that run dissimilar protocols and in many cases support dissimilar functions or applications.

Hierarchical Routing

Routing based on a hierarchical addressing system. For example, IP routing algorithms use IP addresses, which contain network numbers, host numbers, and (possibly) subnet numbers.

High-Speed Serial Interface (HSSI)

A network standard for high-speed (up to 52 Mbps) serial communications over WAN links.

Homologation

Conformity of a product or specification to international standards, such as ITU, VCCI, UL, CS, and TUV. Enables portability across company and international boundaries.

Hop Count

A routing metric used to measure the distance between a source and a destination. A *hop* is the passage of a packet through one router. For example, RIP (Routing Information Protocol) uses hop count as a routing metric. The maximum allowable hop count for RIP is 16.

Host

A computer system that is attached to a network or the Internet. Hosts allow users on client machines to connect and share files or transfer information. Individual users communicate with hosts by using client application programs.

HSSI

See *High-Speed Serial Interface*.

HTML (HyperText Markup Language)

The standard way to mark text documents for publishing on the World Wide Web. HTML is marked-up using "tags" surrounded by brackets.

HTTP (HyperText Transfer Protocol)

The protocol used in transmitting and converting HMTL across the Internet and from one network to another.

Hub

Generally, a term used to describe a device that serves as the center of a star-topology network. In Ethernet/IEEE 802.3 terminology, a hub is an Ethernet multiport repeater, which is sometimes referred to as a *concentrator*. The term is also used to refer to a hardware/software device that contains multiple independent but connected modules of network and internetwork equipment. Hubs can be active (where they repeat signals sent through them) or passive (where they do not repeat but merely split signals sent through them).

Hybrid Network

An internetwork made up of more than one type of network technology, including LANs and WANs.

Hypermedia

The combination of hypertext and multimedia in an online document.

Hypertext

A type of text that allows embedded "links" to other documents. Clicking on or selecting a hypertext link displays another document or section of a document. Most World Wide Web documents contain hypertext.

HyperText Markup Language

See *HTML*.

HyperText Transfer Protocol

See *HTTP*.

I

IANA

See *Internet Assigned Numbers Authority*.

ICMP (Internet Control Message Protocol)

A Network-layer Internet protocol that provides message packages to report errors and other information relevant to IP packet processing. ICMP is documented in RFC 792.

IDN

See *Integrated Digital Network*.

IEEE

See *Institute of Electrical and Electronic Engineers*.

IFIP

See *International Federation for Information Processing*.

In-Band Signaling

Used for transmissions within a frequency range normally used for information transmission. This is in contrast with *out-of-band signaling*, which uses frequencies outside the normal range of information-transfer frequencies.

Institute of Electrical and Electronic Engineers (IEEE)

A professional organization that defines network standards. IEEE LAN standards are the predominant LAN standards today, including protocols similar or virtually equivalent to Ethernet and Token Ring.

Integrated Digital Network (IDN)

The telephone company's digital network. It provides digital service within and between all telephone companies, but not to the customer's door.

Integrated Services Digital Network

See *ISDN*.

Interexchange Carrier

See *IXC*.

Interface

A connection between two systems or devices. In routing terminology, a network connection. Also, the boundary between adjacent layers of the OSI model. In telephony, a shared boundary defined by common physical interconnection characteristics, signal characteristics, and meanings of interchanged signals.

Interior Gateway Protocol (IGP)

An Internet protocol used to exchange routing information within an autonomous system. Examples of common Internet IGPs include IGRP (Internet Gateway Routing Protocol) and RIP (Routing Information Protocol).

Intermediate System (IS)

A routing node in an OSI network.

International Federation For Information Processing (IFIP)

A research organization that performs OSI pre-standardization work. Among other accomplishments, IFIP formalized the original MHS (Message Handling Services) model.

International Organization for Standardization

See *ISO*.

International Standards Organization

See *ISO*.

International Telecommunications Union - Telecommunication Standardization Sector

See *ITU-T*.

Internet

A term meaning "the worldwide network of networks," each connected to each other using the IP protocol suite. The Internet provides file transfer, remote login, e-mail, news, and other services.

Internet Address

See *IP Address*.

Internet Assigned Numbers Authority (IANA)

The central registry for various Internet protocol parameters, such as port, protocol, and enterprise numbers and options, codes, and types. The currently assigned values are listed in the "Assigned Numbers" document. If you'd like more information or want to request a number assignment, you can e-mail IANA at iana@isi.edu.

Internet Control Message Protocol

See *ICMP*.

Internet Information Center

See *InterNIC*.

Internet Protocol

See *IP*.

Internet Protocol Address

See *IP Address*.

Internet Relay Chat (IRC)

The current worldwide "party line," IRC allows multiple users to converse in real time on different "channels." Channels (which have a # sign preceding their name) vary in traffic and content. Channel operators (or "Ops") moderate the conversation and have the ability to "kick" people from channels, or even ban them if their actions warrant it. IRC clients are available for nearly all platforms.

Internet Research Task Force (IRTF)

A community of network researchers with an internetwork focus. The IRTF is governed by the Internet Research Steering Group (IRSG).

Internet Service Provider (ISP)

A company that maintains a network that is linked to the Internet via a dedicated communication line, usually a high-speed link known as a T1. An ISP offers use of its dedicated communication lines to companies or individuals. Using a modem, users can dial up to an ISP whose computers will connect them to the Internet, typically for a fee.

Internetwork

A collection of networks interconnected by routers that functions (generally) as a single network. Sometimes called an *internet*, which is not to be confused with the *Internet* (with a capital *I*). *Internetworking* is a general term used to refer to connecting networks together. The term can describe products, procedures, and technologies.

Internetwork Packet Exchange

See *IPX*.

InterNIC

The combined name for the providers of registration, information, and database services to the Internet. InterNIC is who you contact if you want to register a domain name on the Internet.

Interoperability

The ability of computing equipment manufactured by different vendors to communicate successfully over a network.

IOC

See *ISDN Ordering Code*.

IP (Internet Protocol)

An industry-standard, connectionless, best-effort, packet-switching protocol used as the Network layer in the TCP/IP protocol suite. IP contains addressing information and some control information that allows packets to be routed. IP is documented in RFC 791.

IP Address

A 32-bit address defined by the IP and assigned to hosts using TCP/IP. Every resource on the Internet has a unique numerical IP address, represented in dotted-decimal notation. The address is made up of a network section, an optional subnet section, and a host section. IP addresses are similar to phone numbers. When a user "calls" that number (using any number of connection methods such as FTP, HTTP, Gopher, etc.), that user is connected to the computer that "owns" that individualized IP address.

IPX (Internetwork Packet Exchange)

A Novell Network layer protocol similar to XNS and IP, which is used in NetWare networks.

IRC

See *Internet Relay Chat*.

ISDN (Integrated Services Digital Network)

A relatively new telecommunications technology which combines voice and digital network services in a single medium. ISDN makes it possible for communications carriers to offer their customers digital data services as well as voice connections through a single line. ITU-T defines the standards relating to ISDN.

ISDN Bridge

A LAN-to-LAN connectivity device to link one LAN to another, with dialed connections made on demand over ISDN.

ISDN Ordering Code (IOC)

A predefined code that tells the telephone company how to provision your ISDN line based on the requirements of your ISDN hardware. An example

is "Intel Blue," which is usually used for setting up Intel's videoconferencing desktop systems.

ISO (International Organization for Standardization)

An international organization of 89 member countries (founded in 1946) responsible for setting world standards in many electronics areas, including those relevant to networking. The ISO is responsible for the most popular networking reference model: the OSI reference model.

ISO Development Environment (ISODE)

A popular implementation of OSI's upper layers on a TCP/IP protocol stack.

Isochronous Transmission

Asynchronous (start and stop) transmission over a synchronous data link. In telephony, isochronous implies constant bit-rate sampling and is referred to as the inverse of asynchronous transmission. Voice and video transmissions are examples of isochronous transmission.

ISODE

See *ISO Development Environment*.

ISP

See *Internet Service Provider*.

ITU-T (International Telecommunications Union - Telecommunication Standardization Sector)

An organization that produces technical standards, known as *Recommendations*, for all internationally controlled aspects of analog and digital communications. Formerly known as the CCITT.

IXC (Interexchange Carrier)

A telephone company that handles long-distance connections.

J

Japan Unix Network (JUNET)

The largest nationwide, noncommercial network in Japan, designed to promote communication between Japanese and outside researchers.

Jitter

Analog communication line distortion caused by a variation of a signal from its reference timing positions. Jitter can cause data loss, particularly at high speeds.

John Von Neumann Center Network

A regional network composed of T1 and slower serial links providing mid-level networking services to sites in the Northeast.

JUNET

See *Japan Unix Network*.

K

Keepalive Message

A message sent by one network device to inform another network device that the virtual circuit between the two is still active.

L

LAN (Local-Area Network)

A network covering a relatively small geographic area (usually not larger than a floor or small building), such as Ethernet and Token Ring networks. LANs are now commonplace in most businesses, allowing users to send e-mail and share resources such as files, printers, modems, etc. Some larger companies are connecting their LANs to the Internet, allowing users to connect to resources within or outside the LAN.

LAP-B (Link Access Procedure, Balanced)

Derived from HDLC (High-level Data Link Control), a CCITT X.25 version of a bit-oriented data link protocol.

LAPD (Link Access Procedures-D Channel)

ISDN's Data-Link-layer protocol for the D channel. LAPD was derived from the CCITT X.25 LAP-B protocol and is designed primarily to satisfy the signaling requirements of ISDN Basic Rate Interface. LAPD is defined by CCITT Recommendations Q.920 and Q.921.

LAT

See *Local Area Transport*.

LATA

See *Local Access and Transport Area*.

Latency

The amount of time between when a device requests access to a network and when it is granted permission to transmit.

Leased Line

A dedicated, full-time transmission line or connection. Commonly, a leased line is reserved by a communications carrier for the private use of a customer. A leased line can be used to link a user or network to an Internet service provider or another network.

LEC (Local Exchange Carrier)

A telephone company that handles local connections, often the same as the local telephone company.

Line Conditioning

The use of equipment on leased voice-grade communications channels to improve analog characteristics, thereby allowing higher transmission rates.

Line Driver

An inexpensive amplifier/signal converter that conditions digital signals to ensure reliable transmissions over extended distances.

Line Extension

A means for providing ISDN at distances beyond the normal limit between the central office and your location.

Line Turnaround

The time required to change data transmission direction on a telephone line.

Link

A network communications channel consisting of a circuit or transmission path, including all equipment, between a sender and a receiver. Most often used to refer to a WAN connection. Sometimes referred to as a *line*.

Link Access Procedure, Balanced

See *LAP-B*.

Link Access Protocol D

See *LAP-D*.

Listserv

An automated mailing list distribution system. Listservs exist for a multitude of professional, educational, and special interest groups. Usually, the user sends an e-mail message to a Listserver with the subject "SUBSCRIBE *listname*" or something similar. The user is then "subscribed" to that "mailing list" and (depending on the service) will receive regular mail from a single source or from all members who send e-mail to the Listserver. Listserv was originally designed for the BITNET/EARN network.

Load Balancing

In routing, the ability of a router to distribute traffic over all its network ports that are same distance from the destination address. Good load-balancing algorithms use both line speed and reliability information. Load balancing increases the utilization of network segments, thus increasing effective network bandwidth.

Local Access and Transport Area (LATA)

A telephone dialing area serviced by a single, local telephone company. Calls within LATAs are called local calls. There are well over a hundred LATAs in the United States.

Local Acknowledgment

A method whereby an intermediate network node, such as a router, terminates a Data-Link layer session for an end host. Use of local acknowledgments reduces network overhead and, therefore, the risk of time-outs.

Local-Area Network

See *LAN*.

Local Area Transport (LAT)

A network virtual terminal protocol developed by Digital Equipment Corporation.

Local Bridge

A bridge that directly interconnects networks in the same geographic area.

Local Exchange Carrier

See *LEC*.

Local Loop

The line from a telephone subscriber's premises to the telephone company's central office.

Localtalk

Apple's proprietary 230-kbps, baseband, carrier sense multiple access with collision detection (CSMA/CD) network protocol.

Logical Channel

A nondedicated, packet-switched communications path between two or more network nodes. Through packet switching, many logical channels can exist simultaneously on a single physical channel.

Loop

A route in which packets never reach the destination, but simply cycle repeatedly through a constant series of network nodes.

Loop Qualification

A test that a telephone company runs to make sure that a customer's ISDN line meets the distance and quality requirements of 18,000 feet from the central office that provides the ISDN service. Also known as *line qualification*.

Loopback Test

A test in which signals are sent and then directed back toward the source at some point along the communications path. Loopback tests are often used to test network interface usability.

Lossy

Characteristic of a network that is prone to lose packets when it becomes highly loaded.

M

Mail Reflector

A program that distributes files or information in response to requests sent via e-mail. Many Listservs have mail reflectors. Users can request documents of a reflector by sending a message with the subject "SEND *document name*" or a similar command. Mail reflectors are also being used to provide FTP-like services for users with limited Internet access.

Mailing List

A list of e-mail addresses used to forward messages to groups of people. When users subscribe to a mailing list, they receive all mail sent to that list.

MAN (Metropolitan Area Network)

A network that spans a metropolitan area. Generally, a network that spans a larger geographic area than a LAN, but a smaller geographic area than a WAN.

Maximum Transmission Unit

The maximum packet size, in bytes, that a particular interface will handle.

MCU

See *Multipoint Control Unit*.

Media

In communications, the physical environment through which transmission signals pass. Common network media include twisted-pair, coaxial, and fiber-optic cable, and the atmosphere (through which microwave, laser, and infrared transmission occurs).

Message Handling System

See *MHS*.

Message Switching

A switching technique involving transmission of messages from nodes through a network. The message is stored at each node until a forwarding path is available.

Metropolitan Area Network

See *MAN*.

MHS (Message Handling System)

CCITT X.400 recommendations that provide message handling services for communications between distributed applications. NetWare MHS is a different (though similar) entity that also provides message-handling services and is marketed by Novell.

MIDAS

An Australian public switched telephone network.

Midsplit

A broadband cable system in which the available frequencies are split into two groups: one for transmission and one for reception.

MIME (Multipurpose Internet Mail Extension)

A standardized method for organizing divergent file formats. The method organizes file formats according to the file's MIME type. When Internet software (such as a browser or e-mail program) retrieves a file from a server, the server provides the MIME type of the file, and the file is decoded correctly when transferred to the user's machine.

Mirror Site

Areas on a computer that "mirror" or contain an exact replica of the directory structure of another computer. If you have trouble getting connected to an FTP site, for example, because of the high amount of traffic, you can usually connect to a mirror site that contains the same information on a different computer. Mirror sites are usually updated once a day.

Modem

A device that translates digital pulses from a computer into analog signals for telephone transmission, and analog signals from the telephone into digital pulses the computer can understand. The modem provides communication capabilities between computer equipment over regular telephone facilities.

Modulation

The process by which signal characteristics are transformed to represent information. Types of modulation include frequency modulation (FM), in which signals of different frequencies represent different data values, and amplitude modulation (AM), in which signal amplitude is varied to represent different data values.

Mosaic

A graphical browser for the World Wide Web that supports hypermedia. The NCSA (National SuperComputer Association) invented the Mosaic browser, which quickly became the industry standard. However, other companies, such as Microsoft (Internet Explorer) and Netscape Communications have introduced their own browsers. The term *Mosaic* is sometimes used incorrectly as a synonym for the World Wide Web.

MP (Multichannel Protocol)

Defines a method for equipment to inverse multiplex a packet data stream between two points. It describes how the transmitting point breaks the data stream into smaller individual data streams and sends them out over multiple transmission channels. It also describes how the receiving point accepts the incoming data streams and reassembles them into the original data.

MPPP

See *Multilink PPP*.

MPF (Multilink Packet Fragmentation)

Splitting each packet and sending the parts on both channels.

MUD

See *Multi-User Dungeon*.

Mu-Law

North American standard used in conversion between analog and digital signals in PCM (Pulse Code Modulation) systems. Similar to the European A-law.

Multicast

Single packets copied to a specific subset of network addresses. There addresses are specified in the destination-address field. In contrast, in a broadcast, packets are sent to all devices in a network.

Multicast Address

An address that refers to multiple network devices. Synonymous with group address.

Multilink PPP (MLPPP)

A protocol that allows two or more B channels of an ISDN line to be combined into a single, faster PPP (Point-to Point Protocol) connection. With Multilink PPP, a 128-kbps PPP connection is possible over a Basic Rate ISDN line. MLPPP+ (somtimes referred to as MP+) adds transparent bandwidth-on-demand capabilities to MP (Multichannel Protocol), the Internet standard inverse multiplexing or channel bonding protocol, specified in RFC 1717.

Multiplexing

In communications, a technique for allowing multiple signals to share a single channel.

Multipoint

A communication configuration in which several terminals or stations are connected. This is in contrast to a point-to-point configuration, where communication is between two stations only.

Multipoint Control Unit (MCU)

A device that bridges together multiple inputs so that more than three parties can participate in a videoconference. The MCU uses fast switching techniques to patch the presenter's input to the output ports representing the other participants.

Multipoint Line

A communications line having multiple cable access points. Also called *multidrop line*.

Multipurpose Internet Mail Extension

See *MIME*

Multi-User Domain

See *Multi-User Dungeon*.

Multi-User Dungeon (MUD)

Role-playing games that take place on a computer. Users can Telnet to a MUD host and create a character (and save characters for future sessions). MUDs can be action-based, adventure-based, or fantasy-based games. Some MUDs have thousands of registered characters, and most foster a community or culture of their own.

Multivendor Network

A network using equipment from more than one vendor. Multivendor networks pose many more compatibility problems than single-vendor networks.

N

Name Resolution

Generally, the process of associating a name with a network location.

Name Server

A server provided on the network that resolves network names into network locations (addresses).

Narrowband ISDN (N-ISDN)

Subvoice channels able to carry data starting at speeds up to 64 kbps, ranging up to T1 rates. Sometimes used to refer to regular telephone and nonvideo capable systems.

National ISDN

A technical specification (standard) issued by Bell Research Corporation (Bellcore), which attempts to standardize the ISDN services offered by the seven Regional Bell Operating Companies (the owners of Bellcore). This specification serves as a recommendation to other communications companies in the United States.

National Institute of Standards and Technology (NIST)

Formerly the National Bureau of Standards (NBS), the United States government organization that supports and catalogs a variety of standards.

National ISDN Users Forum (NIUF)

Sponsored by the United States Department of Commerce under the National Institute of Standards and Technology (NIST), the NIUF assists in the development of ISDN applications and educates users on ISDN services.

National Science Foundation Network (NSFN)

A large network controlled by the National Science Foundation that provides networking services in support of education and research in the United States.

NBS (National Bureau of Standards)

See *National Institute of Standards and Technology*.

NDIS

See *Network Driver Interface Specification*.

Network

A collection of computers and other devices that are able to communicate with each other over some network medium.

Network Address

Also called a protocol address, a network layer address referring to a logical, rather than a physical, network device.

Network Driver Interface Specification (NDIS)

Produced by Microsoft, a specification for a generic, hardware-independent and protocol-independent device driver for NICs (network interface cards).

Network File System (NFS)

A file system that allows a computer to access and use files over a network as if they were local. Common use is a distributed file system protocol suite developed by Sun Microsystems that allows remote file access across a network. NFS is simply one protocol in the suite of protocols. NFS protocols include NFS, XDR (External Data Representation), RPC (Remote Procedure Call), and others. These protocols are part of a larger architecture that Sun refers to as ONC (Open Network Computing).

Network Layer

See *OSI Reference Model*.

Network Management System (NMS)

A system responsible for managing at least part of a network. An NMS is generally a reasonably powerful and well-equipped computer such as an engineering workstation, with a megapixel color display, large memory, lots of disk space, and a fast processor. NMSs communicate with agents to help keep track of network statistics and resources.

Network News Transfer Protocol (NNTP)

An industry-standard protocol for the distribution, inquiry, retrieval, and posting of news articles.

NFS

See *Network File System*.

NI-1

Short for National ISDN 1. It is a national standard for ISDN interoperability between switches NI-1, or NI-*x*.

N-ISDN

See *Narrowband ISDN*.

NIST

See *National Institute of Standards and Technology*.

NIUF

See *National ISDN Users Forum*.

NMS

See *Network Management System*.

NNTP

See *Network News Transfer Protocol*.

Node

A generic term used to refer to an entity with a network interface card installed that can access a network. This is usually a computer but can be another device, such as a printer or modem.

Northern Telecom DMS

See *DMS-100*.

Northwest Net

A regional network, founded by the National Science Foundation, serving the Northwest, Alaska, Montana, and North Dakota. Northwest Net connects all major universities in the region, as well as many leading industrial concerns such as Boeing and Sequent Computer.

NPDN

A low-speed, circuit-switched public network in the Nordic countries, including, Denmark, Finland, Iceland, Norway, and Sweden.

NSFNET

See *National Science Foundation Network*.

NT1 (Network Termination 1)

The device that connects to ISDN hardware and works as a converter between an ISDN U interface and an ISDN S/T interface. Some ISDN adapters have a NT1 already built into them. This is easier and less expensive than an external NT1, but may prevent you from connecting other equipment to your ISDN line.

Null Modem

A small box or cable used to join computing devices directly, rather than over a network.

Numeris

A public ISDN network in France.

NYSERNET

New York State network with a T1 backbone connecting the National Science Foundation, many universities, and several commercial concerns.

OARNET

Ohio Academic Resources Network. Connects sites, including the Ohio Supercomputer center in Columbus.

Office Document Architecture

An OSI standard that specifies how documents are transmitted electronically.

ONC

See *Open Network Computing*.

Open Architecture

An architecture according to which third-party developers can legally develop products and for which public domain specifications exist.

Open Circuit

A broken path along a transmission medium. Open circuits will usually prevent network communication.

Open Network Computing (ONC)

A distributed applications architecture founded by Sun Microsystems, currently controlled by a consortium led by Sun. The NFS (Network File System) protocols are part of ONC.

Open System Interconnection (OSI)

An international standardization program created by ISO and ITU to develop standards or a suite of protocols for data networking. OSI facilitates multivendor equipment interoperability.

Ordering Code

See *ISDN Ordering Code.*

OSI

See *Open System Interconnection.*

OSI Reference Model

A network architectural model developed by the ISO and CCITT. The model consists of seven layers, each of which specifies particular network functions such as addressing, flow control, error control, encapsulation, and reliable message transfer. The layers are Physical, Data-Link, Network, Transport, Session, Presentation, and Application. The highest layer (the Application layer) is closest to the user; the lowest layer (the Physical layer) is closest to the media technology. The OSI reference model is used universally as a method for teaching and understanding network functionality.

Out-of-Band Signaling

Transmission outside a frequency range normally used for information transmission. This is in contrast with *in-band signaling*, which uses frequencies inside the normal range of information-transfer frequencies.

P

Packet

The common term for the standard unit of data sent across a network. This logical grouping of information includes a header and (usually) user data or a trailer.

Packet Buffer

A storage area to hold incoming data until the receiving device can process the data.

Packet-Switch Node (PSN)

An Internet packet switch. Also, a switching node in the X.25 architecture. Usually, the PSN is data communications equipment (DCE) and allows for connection to data terminal equipment (DTE).

Packet-Switched Data (PSD)

Data that is broken down into standard-sized packets, then sent through the network. The packets each carry the destination address and the information required to reassemble them when they reach their destination.

Packet-Switched Network (PSN)

A method of digital communication in which messages are divided into packets of a bit size determined by the needs of the transmission network, and are transferred to their destination over communication channels that are dedicated to the connection only for the duration of the packet's transmission. This method allows nodes to share bandwidth with each other by intermittently sending logical information units (packets). In contrast, a circuit-switching network dedicates one circuit at a time to data transmission.

PAL

TV standard used in Britain, Germany, and most of the rest of western Europe excluding France. It's similar to the American system (NTSC), with the main difference indicated by its name, which stands for phase alternate line (colloquially, "perfection at last"). The method for encoding hue is reversed between lines. Thus, if some error causes the even-numbered lines to be too green, the same error will cause odd-numbered lines to be too magenta. The human eye averages the lines together, and one sees accurate hues in spite of the error.

PAP (Password Authentication Protocol)

One of the many authentication methods that can be used when connecting to an Internet service provider. PAP allows you to log in automatically, without using a terminal window to type in your username and password. One warning about PAP: passwords are sent over the connection in text format, which means there is no protection if someone is "listening-in" on your connection.

Parc Universal Protocol

See *PUP*.

Password Authentication Protocol

See *PAP*.

Path Cost

An arbitrary value used as a routing metric to determine the best path for sending data to a network destination.

Payload

The information content of a transmission system, such as the 192 bits of information in a DS-1 frame.

PBX (Private Branch Exchange)

A private telephone switch that provides switching (including a full set of switching features) for an office or campus. PBXs often use proprietary digital-line protocols, although some are analog-based.

PCM

See *Pulse Code Modulation*.

PDN

See *Public Data Network*.

Peer-to-Peer Computing

A form of computing in which each network device runs both the client and server portions of an application. The phrase can also be used to describe communication between implementations of the same OSI reference model layer in two different network devices.

Physical Address

A term sometimes used to refer to the Data-Link-layer address of a network device. Contrasts with a network or protocol address, which is a Network-layer address. Also called the *hardware address*.

Physical Layer

See *OSI Reference Model*.

Piggybacking

The process of carrying acknowledgments within a data packet to save network bandwidth.

Ping

The simplest way to test or time the response of an Internet connection. Ping sends a request to an Internet host and waits for a reply. Appropriately, the response is called a Pong. When you Ping an address, you get a response telling you the number of seconds it took to make the connection. Ping clients exist for a number of platforms, or you can use a Unix or Windows 95 prompt to issue a Ping command directly.

Ping-Ponging

A phrase used to describe the actions of a packet in a two-node routing loop.

Point-to-Point Protocol

See *PPP*.

POP

See *Post Office Protocol*.

Port

An access point to a computer, peripheral, network, circuit switch, or other device (*hardware port*). A parallel port is a hardware connection where there are separate pins defined for all 8 data bits in a character, so that an entire byte of information can be sent at a time. A serial port is one in which only one pin is available for data transmission in a given direction, so that bits must be transmitted in sequence. A *software port* is a memory location that is in association with a hardware port or a communications channel.

Also, an emerging term, somewhat slang, for an installation of telecommunications equipment, usually digital leased lines and multiprotocol routers.

Post Office Protocol (POP)

A protocol designed to allow single users to read mail from a server. There are three versions: POP, POP2, and POP3. When e-mail is sent to a user, it is stored on the server until accessed the user accesses it. Once the user is authenticated, the POP is used to transmit the stored mail from the server to your local mailbox on your client machine.

Postal Telephone and Telegraph

See *PTT*.

Posting

Sending an article to a Usenet newsgroup or placing a message on a bulletin board service (BBS).

POTS (Plain Old Telephone Service)

The standard analog telephone service used by many telephone companies throughout the United States.

Presentation Layer

See *OSI Reference Model*.

PPP (Point-to-Point Protocol)

A protocol that provides a method for transmitting packets over serial point-to-point links. PPP is one of the most popular methods for dial-up connections to the Internet, since it allows the use of other standard protocols (such as IPX, TCP/IP, and NetBUI) over a standard telephone connection, but it can also be used for LAN connections.

PRI

See *Primary Rate Interface*.

Primary Rate Interface (PRI)

One of two subscriber interfaces for ISDN (the other is Basic Rate Interface). PRI consists of a single 64-kbps D channel plus twenty-three 1.544-Mbps B channels or thirty 2.048-Mbps B channels for voice and/or data. This is the ISDN equivalent of a T1 line delivered to the customer's premise. Also known as 23B+D.

Private Branch Exchange

See *PBX*.

Propagation Delay

The time required for data to travel over a network from its source to its ultimate destination.

Protocol

The "language" spoken between computers to help them exchange information. A protocol is a formal description of message formats and the rules that two computers must follow to exchange those messages. Protocols can describe low-level details of machine-to-machine interfaces (like the order in which bits and bytes are sent across a wire) or high-level exchanges between allocation programs (the way in which two programs transfer a file across the Internet). Standard protocols allow different manufacturers' computers to communicate. The different equipment and computers can use completely different software, provided each computer's software can agree upon the meaning of the data.

Protocol Converter

A device that enables equipment with different data formats to communicate by translating the data transmission code of one device to the data transmission code of another device. Also known as a *protocol translator*.

Protocol Stack

Related layers of protocol software that function together to implement a particular communications architecture. Examples include Appletalk and DECnet.

Provisioning

The combination of device and service options that make up a customer's ISDN line. The customer orders the ISDN line, but the telephone company provisions the line. The telephone company configures the ISDN service according to the physical capabilities of the switch, as well as the options the customer chose.

PSD

See *Packet-Switched Data*.

PSN

See *Packet Switch Node; Packet-Switched Network*.

PSTN

See *Public Switched Telephone Network*.

PTT (Postal Telephone and Telegraph)

A government agency that provides telephone services. PTTs exist in most areas outside North America and provide both local and long-distance telephone services.

Public Data Network (PDN)

A network operated either by a government (as in Europe) or by a private concern to provide computer communications to the public, usually for a fee. PDNs enable small organizations to create a WAN without all the equipment costs of long-distance circuits.

Public Switched Telephone Network (PSTN)

A public network that provides circuit-switching telephone service to users.

Pulse Code Modulation (PCM)

Transmission of analog information in digital form through sampling and encoding the samples with a fixed number of bits.

PUP (PARC Universal Protocol)

A protocol similar to IP (Internet Protocol) developed at Xerox Palo Alto Research Center.

Q

Q.920/Q.921

ISDN specifications for the user-network interface at the Data-Link layer of the OSI model.

Q.931

A CCITT recommendation that is a standard for signaling to set up ISDN connections.

Q2931

CCITT recommendation that is a standard for signaling to set up ATM virtual connections. An evolution of CCITT Recommendation Q.931.

Quality of Service (QOS)

A measure of performance for a transmission system that reflects its transmission quality and availability of service.

Queue

Generally, an ordered list of elements waiting to be processed. In routing, a backlog of packets waiting to be forwarded over a router interface.

Queuing Delay

The amount of time that data must wait before it can be transmitted onto a statistically multiplexed physical circuit.

R

RACE

See *Research and Development Program in Advanced Communications in Europe*.

Radio Austria

A public service telephone network based in Austria.

Radio Frequency (RF)

Generic term referring to frequencies that correspond to radio transmissions. Cable television (CATV) and broadband networks use RF technology.

RARE

See *Reseaux Associes pour la Recherche Europeenne*.

RBOC (Regional Bell Operating Company)

The seven holding companies formed by the divestiture of the former AT&T Bell System. The RBOCs are Ameritech, Bell Atlantic, Bell South, NYNEX, Pacific Bell (or Pacific Telesis), SBC Communications, and US West. RBOCs are also known as RHCs (Regional Holding Companies), RBHCs (Regional Bell Holding Companies), and more generally as LECs (local-exchange carriers).

Real-Time

The processing of information that returns a result so rapidly that the interaction appears to be instantaneous. Telephone calls and videoconferencing are examples of real-time applications.

Redundancy

In telephony, the portion of the total information contained in a message that can be eliminated without loss of essential information or meaning. In computing, multiple (redundant) system elements that perform the same function.

Regional Bell Holding Company

See RBOC.

Regional Bell Operating Company

See RBOC.

Relay

OSI terminology for a device that connects two or more networks or network systems. A Layer 2 relay is a bridge; a Layer 3 relay is a router.

Remote Bridge

A bridge that connects physically disparate network segments via WAN links.

Repeater

A device that regenerates and propagates electrical signals between two network segments.

Request for Comments

See *RFC*.

Research and Development Program in Advanced Communications in Europe

A project sponsored by the European Community for the development of broadband networking capabilities.

Reseaux Associes Pour La Recherche Europeenne

An association of European universities and research centers designed to promote an advanced telecommunications infrastructure in the European scientific community.

RF

See *Radio Frequency*.

RFC (Request for Comments)

In the Internet community, a document series that describes the Internet suite of protocols and related experiments. The documents are the primary means for communicating information about the Internet. Most RFCs document protocol specifications such as Telnet and FTP, but some are humorous and/or historical. RFCs are available from the Internet Network's Information Centers (InterNIC).

RGB

For Red, Green, Blue, the additive used in color video systems. Color television signals are oriented as three separate pictures: red, green, and blue. Typically, they are merged together as a composite signal, but for maximum quality and in computer applications, the signals are segregated.

Ring Topology

Topology in which the network consists of a series of repeaters connected to one another by unidirectional transmission links to form a signal closed loop. Each station on the network connects to the network as a repeater.

RIP (Routing Information Protocol)

The most common IGP (Interior Gateway Protocol) in the Internet. RIP uses hop count as a routing metric. The largest allowable hop count for RIP is 16.

RJ-11

The most common telephone jack in the world, this is a six-conductor modular jack with four wires. Almost everyone has RJ-11 jacks in their house.

RJ-45

An eight-pin connector jack used with standard telephone lines and required by some ISDN hardware. An RJ-45 is a little larger than an RJ-11 jack.

Routed Protocol

A protocol that can be routed by a router. To route a routed protocol, a router must understand the logical internetwork as perceived by that routed protocol. Examples of routed protocols include DECnet, Appletalk, and IP.

Router

A software and hardware system that connects two or more networks sometimes called a gateway (although this definition of gateway is becoming increasingly outdated). A router is responsible for making decisions about which of several paths will be followed by network (or internetwork) traffic. A router determines where the destination computer is located, and then finds the best way to get there, often using routing algorithms to choose the best route.

Routing

The process of finding a path to the destination host. Routing is very complex in large networks because of the many potential intermediate destinations a packet might traverse before reaching its destination host.

Routing Information Protocol

See *RIP*.

Routing Metric

The method by which a routing algorithm determines that one route is better than another. This information is stored in routing tables. Metrics include reliability, delay, bandwidth, load, maximum transmission units, communication costs, and hop count.

RS-232-C

A popular Physical layer interface. Virtually identical to the V.24 specification. Now EIA-232-D.

RS-422

A balanced electrical implementation of RS-449 for high-speed data transmission. Now EIA-422.

RS-423

An unbalanced electrical implementation of RS-449 for RS-232-C compatibility. Now EIA-423.

RS-449

A popular Physical layer interface. Essentially a faster (up to 2 Mbps) version of RS-232-C capable of longer cable runs. Now EIA-449.

S

S/T Interface

A six-wire interface. All traffic on the S/T bus flows in 48-bit frames, at a transmission rate of 192 kbps. This is a higher rate than the 128 kbps that is sent between the customer and the phone company (64+64+16), because the customer premise installation covers shorter distances. A rate of 144 kbps is used for the 2B+D channels, leaving 48 kbps for the overhead. Since the S/T bus needs to handle network contention (two signals wanting the same channel) in addition to other issues, it needs all of this extra bandwidth to keep things running smoothly.

SAPONET-P

South African public packet-switching data network.

SDLC (Synchronous Data Link Control)

IBM bit-synchronous Data Link-layer protocol that has spawned numerous similar protocols, including HDLC and LAP-B.

SDLC Transport

A router feature that allows disparate environments to be integrated into a single, high-speed, enterprise-wide network. A router with this feature can pass native SDLC traffic through point-to-point serial links and multiplex other protocol traffic over the same links.

SECAM (Sequential Couleur Avec Memoire)

The color television system offering 625 scan lines and 25 interlaced frames per second. It was developed after NTSC and PAL and is used in France, the former Soviet Union, the Former Eastern bloc countries, and parts of the Middle East. Two versions of SECAM exist: horizontal SECAM and vertical SECAM. In this system, frequency modulation is used to encode the chrominance signal.

Serial Line Internet Protocol

See *SLIP*.

Serial Transmission

A method of data transmission in which the bits of a data character are transmitted sequentially over a single channel.

Server

A node or software program that provides services to a client. In other words, a computer that provides resources, such as files or other information. Common Internet servers include file servers and DNS servers.

Service Profile Identifier

See *SPID*.

Session Layer

See *OSI Reference Model*.

Shielded Cable

Cable that has a layer of shielded insulation to reduce electromagnetic interference.

Signaling

In telephony, refers to the way that information pertaining to call setup is passed from node to node in the telephone system. The most commonly known form of signaling is the frequencies associated with touch-tone. These frequencies are the telephone system's way of supplying the central office with the information that it requires to process a call. With ISDN, all signaling is done on a separate data circuit (out-of-band), using a standard known as Signaling System 7 (SS7).

Signaling System 7 (SS7)

A standard used with ISDN to perform signaling. SS7 provides out-of-band signaling for call processing and routing. This allows for the exchange of caller ID numbers and call routing prior to call connection or rejection.

Simple Mail Transfer Protocol

See *SMTP*.

Simple Network Management Protocol

See *SNMP*.

Simplex Transmission

Data transmission in only one direction.

S Interface

An interface for Basic Rate ISDN delivered over four wires. The maximum loop length for this interface is one kilometer (1.6 miles).

SLC (Subscriber Line Charge)

See *Customer Access Line Charge*.

SLIP (Serial Line Internet Protocol)

Similar to PPP, SLIP is another standard protocol used to run TCP/IP over serial lines, such as telephone circuits or RS-232 cables. Unlike PPP, however, SLIP does not work on LAN connections.

SMDS (Switched Multimegabit Digital Service)

A high-speed, packet-switched, datagram-based, WAN technology offered by the telephone companies.

SMTP (Simple Mail Transfer Protocol)

An Internet protocol providing e-mail services. SMTP transfers mail from server to server. End users must use POP (Post Office Protocol) to transfer the messages to their machines.

SNA (Systems Network Architecture)

A large, complex, feature-rich network architecture developed in the 1970s by IBM.

SNAP (Subnetwork Access Protocol)

An Internet protocol that operates between a network entity in the subnetwork and a network entity in the end system and specifies a standard method of encapsulating IP datagrams and ARP messages on IEEE networks. The SNAP entity in the end system makes use of the services of the subnetwork and performs three key functions: data transfer, connection management, and quality of service selection.

SNMP (Simple Network Management Protocol)

A protocol developed to manage nodes on an IP network, SNMP is an Internet standard protocol. It can be used to manage wiring hubs, video toasters (units for managing video via computer), CD-ROM jukeboxes, and many other devices. SNMP provides a means to monitor and set network configuration and runtime parameters.

SONET (Synchronous Optical Network)

A high-speed (up to 2.5 Gbps) synchronous network approved as an international standard in 1988. The RBOCs are likely to make SONET popular as a transmission system underlying SMDS. Now called SDH (Synchronous Digital Hierarchy).

Source-Route Bridging

A method of bridging originated by IBM where the entire route to a destination is predetermined, in real time, prior to the sending of data to the destination. Contrast this with transparent bridging, wherein bridging occurs on a hop-by-hop basis.

Source-route bridging (sometimes abbreviated to SRB) is most popular in Token Ring networks.

Space Physics Analysis Network (SPAN)

A data comparison network serving NASA projects and facilities, with extensions to Japan, Canada, and many European countries.

Span

Refers to a full-duplex digital transmission line between two digital facilities.

SPID (Service Profile Identifier)

A number that uniquely identifies the device on an ISDN line so that traffic may be routed to it and also so that the capabilities of the ISDN device may be identified.

SS7

See *Signaling System 7*.

Star Topology

A LAN topology in which end points on a network are connected to a common central switch by point-to-point links.

Store and Forward

A message-switching technique in which messages are temporarily stored at intermediate point between the source and destination until such time as network resources (such as an unused link) are available for message forwarding.

Subchannel

In broadband terminology, a frequency-based subdivision creating a separate communications channel.

Subnet Mask

See *Address Mask*.

Subnetwork

Term sometimes used to refer to a network segment. In IP networks, a network sharing a particular subnet address.

Subnetwork Access Protocol

See *SNAP*.

Switch

A device that forwards electrical signals from one cable or LAN segment to another based upon the address information found in the address header of a packet.

Switched 56

A service that allows customers to dial up and transmit digital information up to 56,000 bits per second in much the same way that they dial up an analog telephone call. The service is billed as a monthly charge plus a cost for each minute of usage. Nearly all local and long-distance telephone companies offer Switched 56 service, and any Switched 56 service can connect with any other service, regardless of which carrier offers the service.

Switched Digital

A dial-up transmission service over digital lines such as ISDN.

Switched Multimegabit Data Service

See *SMDS*.

Synchronous Data Link Control

See *SDLC*.

Synchronous Optical Network

See *SONET*.

Synchronous Transmission

A method by which data is transmitted from one communication device to another. A timer at one end of the communication device tells the device to release the data at a preset time interval, thus synchronizing the transmission of data between two points.

Systems Network Architecture

See *SNA*.

T

T Interface

The part of the ISDN line that comes out of the NT1 headed for the computer. Basically, Basic Rate Interface ISDN delivered over four wires.

T.120

A standard for audiographics exchange. T.120 supports higher resolutions than H.320, as well as pointing and annotation.

T1

In Bell system terminology, refers to a digital carrier facility used for transmission of data through the telephone hierarchy. The rate of transmission is 1.544 Mbps.

T3

A digital WAN service used for transmission of data through the telephone hierarchy. The rate of transmission is 44.746 Mbps. A T3 carrier can

handle up to 28 multiplexed T1s or 672 multiplexed 64-kbps digital voice or data channels.

TA

See *Terminal Adapter*.

TAC

See *Terminal Access Controller*.

Talking Head

The portion of a person that can be seen in the typical business-meeting style videoconference: the head and shoulders. This type of image is fairly easy to capture with compressed video because there is very little motion.

T-Carrier

A time-division-multiplexed transmission method usually referring to a line or cable carrying a DS-1 signal.

TCP/IP (Transmission Control Protocol/Internet Protocol)

The two best-known Internet protocols, often erroneously thought of as one protocol. TCP corresponds to Layer 4 (the Transport layer) of the OSI reference model. It provides reliable transmission of data. IP corresponds to Layer 3 (the Network layer) of the OSI reference model and provides connectionless datagram service. TCP/IP is the standard communications protocol set required for Internet computers. To communicate using TCP/IP, PCs need a set of software components called a *TCP/IP stack*. Macintoshes typically use a proprietary software called MacTCP. Most Unix systems are built with TCP/IP capabilities.

TCP/IP Stack

To properly use the TCP/IP protocol, PCs require a TCP/IP stack. This consists of TCP/IP software, sockets software (such as WINSOCK.DLL for Windows machines), and hardware driver software (known as packet drivers). Windows 95 comes with Microsoft's own built-in TCP/IP stack, including version 1.1 of Microsoft's WINSOCK.DLL and packet drivers.

TDM

See *Time Division Multiplexing.*

TDR

See *Time Domain Reflectometer.*

TEI

See *Terminal Endpoint Identifier.*

Telecommunications

Term referring to telephone communications (usually involving computer systems) over the telephone network.

TELENET

A major public service telephone network in the United States.

Telnet

The Internet standard terminal-emulation protocol to connect to remote terminals. Telnet clients are available for most platforms. When a user uses Telnet to connect to a Unix site, for example, that user can issue commands at the prompt as if the machine were local.

Terminal Access Controller (TAC)

An Internet host that accepts terminal connections from dial-up lines.

Terminal Adapter (TA)

The ISDN hardware with a serial data interface. Commonly recognized as the device that connects an ISDN line to a computer, analogous to a modem (except it is digital). The TA does not perform the analog-to-digital and digital-to-analog conversion that a modem does. Instead, the device enables your computer to communicate over digital telecommunications lines and facilities.

Terminal Emulation

A popular network application in which a computer runs software that makes it appear to a host as a directly attached dumb terminal.

Terminal Endpoint Identifier (TEI)

The profile that the central office switch has for the type of terminal that is on the end of the ISDN line.

Terminator

Electrical resistance at the end of a transmission line that absorbs signals on the line, thereby keeping them from bouncing back and being heard again by network stations.

Throughput

The rate of information arriving at, and possibly passing through, a particular point in a network system.

Time Division Multiplexing (TDM)

A technique where information from multiple channels can be allocated bandwidth on a single wire based on time-slot assignment.

Time Domain Reflectometer (TDR)

A device capable of sending signals through a network medium to check cable continuity and other attributes. TDRs are used to find Physical layer network problems.

Timeout

An event that occurs when one network device expects to hear from, but does not hear from, another network device within a specified period of time. The timeout usually results in a retransmission of information or the dissolution of the virtual circuit between the two devices.

TN3270

A slight variation of Telnet used to connect the user to an IBM mainframe. This terminal emulation software that allows a terminal to appear to an IBM host as a 3278 Model 2 terminal. TN3270 clients exist for most platforms.

Token Bus

A LAN architecture using token passing access over a bus topology. This LAN architecture is the basis for the IEEE 802.4 LAN specification.

Token Passing

An access method by which network devices access the physical medium in an orderly fashion based on possession of a small frame called a *token*.

Token Ring

A type of LAN in which networked computers are wired into a "ring." Each computer (or node) is in constant contact with the next node in the ring. A control message, called a *token*, is passed from one node to another, allowing the node with the token to send a message out to the network. If the ring is "broken" by one computer losing contact, the network can no longer communicate. Developed and supported by IBM, the IEEE 802.5 Token Ring standard is the most common.

Topology

The layout or physical arrangement of network nodes and media within an enterprise's networking structure. Topology includes all of the computers on a network and the links that join them.

TOS

See *Type of Service*.

Trailer

Control information appended to the data in a packet.

Transaction

A result-oriented unit of communication processing.

Transmission Control Protocol/Internet Protocol

See *TCP/IP*.

TRANSPAC

A major packet data network run by France Telecom.

Transparent Bridging

A bridging scheme preferred by Ethernet and IEEE 802.3 networks in which bridges pass frames along one hop at a time based on tables associating end nodes with bridge ports. Transparent bridging is so named because the presence of bridges is transparent to network end nodes.

Transport Layer

See *OSI Reference Model*.

Tree Topology

A LAN topology similar to a bus topology, except that tree networks can contain branches. Like the bus topology, transmissions from a station propagate the length of the medium and are received by all other stations.

Trunk

A transmission channel connecting two switching devices.

Twisted-Pair Cable

Another term for regular telephone wiring. Each telephone "wire" is actually a pair of wires. Twisted-pair cable is a relatively low-speed transmission medium consisting of two insulated wires arranged in a regular spiral pattern. The wires may be shielded or unshielded. Twisted-pair cable is very common in telephony applications and is increasingly common in data networks.

TYMNET

A major public service network in the United States.

Type of Service (TOS)

This refers to data or voice and business or residence classifications.

Type of Service Routing

A routing scheme in which the choice of a path through the internetwork depends on the characteristics of the subnetworks involved and the packet, as well as the shortest path to the destination.

U

U Interface

The interface that comes out of the wall in most ISDN installations. This interface usually needs an NT1 before ISDN devices can be attached to it. However, many ISDN modem makers are designing their ISDN devices with built-in NT1s, so that they may hook directly to a U interface. This interface works at loop lengths up to 18,000 feet.

UltraNet

A very high-speed (125-Mbps) network developed and marketed by Ultra Network Technologies Company.

Universal Resource Locator

See *URL*.

Upload

To send a file stored in your machine's storage to another computer. Communications software that facilitates file transfer can be used to transfer and send information, programs, and instructions from the local workstations or desktop computers to a central or larger computer.

Upstream

In data transmission, the communication to the data source. *Downstream* refers to communication from the data source.

URL (Universal Resource Locator)

Refers to the entire address that is recognized "universally" as the address for an Internet resource. Each resource on the Internet has a unique URL. URLs begin with letters that identify the resource type, such as HTTP, FTP, Gopher, etc. These types are followed by a colon and two slashes. Next, the computer's name is listed, followed by the directory and filename of the remote resource.

Usenet

Initiated in 1979, one of the oldest and largest cooperative networks, with more than 10,000 hosts and a quarter of a million users. Its primary service is news, a distributed conferencing service. Usenet groups are more commonly known as *newsgroups*. There are thousands of groups hosted on hundreds of servers around the world, dealing with various topics. Newsreader software is required to properly download and view "articles" in the groups, but people usually "post" an article to a group simply by e-mailing to it.

UUCP

Originally, a program that allowed Unix systems to transfer files over telephone lines. Currently, the term is used to describe the protocol that passes news and e-mail across the Internet.

V

Vendors ISDN Association (VIA)

A nonprofit, California-based association dedicated to making ISDN more accessible to businesses and individual users. ISDN equipment vendors who are members of VIA include Adtran, Ascend Communications, Bay Networks, Cisco Systems, Digi International, Eicon Technology, Intel, Microsoft, Motorola, Shiva, 3Com, and US Robotics.

Veronica

A search engine (not unlike Archie) that is built into Gopher. It allows searches of all Gopher sites for files, directories, and other resources.

VIA

See *Vendors ISDN Association*.

Videoconferencing

The use of video images and audio signals transmitted over communications channels for the purpose of holding meetings when the participants are at different remote sites.

Voice Communications

The transmission of sound in the human hearing range. Voice or audio sound can be transmitted either as analog or digital signals.

W

WAIS

See *Wide Area Information Service*.

WAN (Wide-Area Network)

A network spanning a wide geographic area, usually far enough apart to require telephone communications. WANs typically link cities, and may be owned by a private corporation or by a public telephone company operator. WANs include commercial or educational dial-up networks, such as CompuServe and BITNET.

Web Browser

See *Browser*.

White Pages

Databases containing e-mail addresses, telephone numbers, and postal addresses of Internet users. People can search the Internet White Pages to find information about particular users.

Whiteboarding

A term used to describe the placement of shared documents on an on-screen shared notebook, or "whiteboard." Desktop videoconferencing software includes snapshot tools that enable users to capture entire windows or portions of windows and place them on the whiteboard. Users can also use familiar Windows operations (cut and paste) to put snapshots on the whiteboard.

WHOIS

An Internet program (related to Finger and White Pages) that lets users enter an Internet entity (such as domains, networks, and hosts) and display information such as a person's company name, address, telephone number, and e-mail address.

Wide Area Information Service (WAIS)

A distributed information service and search engine that allows natural language input and indexed searching. Many Web search utilities use a WAIS engine.

Wide Area Network

See *WAN*.

Wideband

Refers to network services that are larger than the T1 (1.544 Mbps) bandwidth.

WinISDN

A Windows ISDN protocol or software interface for PC environments. Used in place of Winsock.

Winsock

For Windows Socket, a set of specifications or standards for programmers creating TCP/IP applications for use with Windows.

World Wide Web (WWW or W3)

A collection of online documents housed on Internet servers around the world. Web documents are written or "coded" in HTML. To access these documents, users run a *Web browser*, such as Netscape or Mosaic. When these browsers access (or hit) a page, the server uses HTTP to send the document to the user's computer.

X

X Windows

A distributed, network-transparent, device-independent, multitasking, windowing and graphics system originally developed by MIT for communication between X terminals and Unix workstations.

XDSL

Refers to the family of services based on ADSL (Asymmetrical Digital Subscriber Line). They include ADSL, HDSL (High-bit rate), VDSL, RADSL, and SDSL.

X.25

An ITU standard that defines the interface for data transfers in a public packet-switched data network. Many establishments have X.25 networks in place that provide remote terminal access. These networks can be used for other types of data, including IP, DECnet, and XNS.

INDEX

Note to the Reader: Throughout this index **boldface** page numbers indicate primary discussions of a topic. *Italicized* page numbers indicate illustrations.

NUMBERS

A

B

(

D

E

M

N

O

P

Q

R

S

T

U

V

ISP Information
Configuration information

☐   Dial-up number: _____

☐   Login name: _____
☐   Login password: _____
☐   E-mail address: _____

☐   Assigned IP number: _____ • _____ • _____ • _____
☐   Gateway IP address: _____ • _____ • _____ • _____

☐   DNS IP address: _____ • _____ • _____ • _____
☐   Subnet mask: _____ • _____ • _____ • _____

☐   POP3 information
   ☐   Mail login: _____
   ☐   Mail password: _____
   ☐   Mail server: _____

☐   SMTP mail server _____

Additional contacts: _____

_____

_____

_____

_____

_____

Notes: _____

_____

_____

_____

_____